The Jaguar XK Series

Other Titles in the Crowood AutoClassics Series

AC Cobra	Brian Laban
Alfa Romeo: Spider, Alfasud and Alfetta GT	David G. Styles
Aston Martin: DB4, DB5 and DB6	Jonathan Wood
Aston Martin and Lagonda V-engined Cars	David G. Styles
Austin Healey 100 and 3000 Series	Graham Robson
BMW M-Series	Alan Henry
Ferrari Dino	Anthony Curtis
Ford Capri	Mike Taylor
Jaguar E-Type	Jonathan Wood
Jaguar Mk 1 and 2	James Taylor
Jaguar XJ Series	Graham Robson
Jensen Interceptor	John Tipler
Lamborghini Countach	Peter Dron
Lotus and Caterham Seven	John Tipler
Lotus Elan	Mike Taylor
Lotus Esprit	Jeremy Walton
Mercedes SL Series	Brian Laban
MGB	Brian Laban
Morgan: The Cars and the Factory	John Tipler
Porsche 911	David Vivian
Porsche 924/928/944/968	David Vivian
Range Rover	James Taylor and Nick Dimbleby
Sprites and Midgets	Anders Ditlev Clausager
Triumph TRs	Graham Robson
Triumph 2000 and 2.5PI	Graham Robson
TVR: The Complete Story	John Tipler

Jaguar XK Series
THE COMPLETE STORY

Jeremy Boyce

First published in 1996 by
The Crowood Press Ltd
Ramsbury, Marlborough
Wiltshire SN8 2HR

© Jeremy Boyce 1996

All rights reserved. No part of this publication may be reproduced or transmitted in any form or by any means, electronic or mechanical, including photocopy, recording, or any information storage or retrieval system, without permisssion in writing from the publishers.

British Library Cataloguing-in-Publication Data

A catalogue record for this book is available from the British Library.

ISBN 1 85223 934 4

Dedication
To Berenice

Picture Credits
All photographs supplied by LAT Photographic, *Autocar*, BMW Great Britain Ltd, The Motoring Picture Library at Beaulieu, Rhoddy Harvey-Bailey, Baxter Bullock and the author.

Printed and bound in Great Britain at BPC Consumer Books, Aylesbury

Contents

	Acknowledgements	6
	Introduction	7
1	The Early Days	8
2	Developing the XK Engine	17
3	The XK Leaps In	28
4	Stretching its Legs – The XK 140	49
5	More Comfort, More Power – The XK 150	61
6	Racing – 1949 and 1950	79
7	Racing – 1951 and Beyond	99
8	Rallying and Record Breaking	122
9	Originality, Buying, Driving and Owning	146
	Appendix I: Production History	173
	Appendix II: Contemporary Road Reports	182
	Index	189

Acknowledgements

My first recollection of a Jaguar is my first recollection of an XK, for I had a pale green Dinky model of the XK 120 fixed-head coupé. I can remember the moment my elder brother informed me that I had a model of the 'fastest car in the world'. This was in the mid-1950s and about the same time a white XK 120 roadster with twin exhausts would very often roar out of the roundabout next to my school and off down the Chertsey Road. My passion for Jaguars grew from there, although I am afraid it has been a mainly nostalgic one in recent years.

In writing this book I built up a list of people who have very generously given their help in many different ways. I wish to thank the following: Peter Sanders, Keith Hollyfield, Andrew Mirylees, Ken Wilks, Julian Kember, Dr Paul Nieuwenhuis, Charles Atkinson, Wilma Keeley, Andy Burnett, Dick Powell, Rhoddy Harvey-Bailey, Graham Hall of XK Engineering, Kathy Ager of LAT Photographic, Ann Harris of Jaguar Daimler Heritage Trust, Quadrant Motor Trader, Brian Davis, Chris Searle and Eric Clow. I am more than grateful to Baxter Bullock who gave his assistance on many, many occasions, all with good humour; and to my wife Petra who has patiently awaited the return of her dishwasher for the very many weeks that it has been toiling on this book.

Afflicted as we are in Britain by an absurd mixture of metric and imperial measurements – almost twenty-five years after our coinage was decimalized – I should like to have used only metric measurements throughout the text. However, since the age of the subject means that imperial measurements are of intrinsic importance to the story, both imperial and metric are given. It was an XK *120* after all.

Introduction

Every so often motor magazines run questionnaires – sometimes in-house, sometimes by asking their readers – to determine the most beautiful cars ever made. The Jaguar XK 120 is usually somewhere near the top. Considering that it was also the fastest volume-production car in the world and in its day won more races than any other sports car, it is without doubt one of those very special cars that arrive only once in a while. Because of this, together with its being considerably cheaper to buy than any rivals it might conceivably have had, as well as cheaper than many cars of greatly inferior performance, it is no wonder that the XK 120 and its derivatives are still viewed with a respect that approaches awe.

It became increasingly obvious during the research for this book that William Lyons was an utterly exceptional man. The remarkably brief period that he took to lead his company from refurbishing Austin Sevens to offering a full range of world-beating performance cars, all of which he styled himself, suggests he might have been a genius.

It has often been pointed out how Lotus has dwindled since the death of its creator and it is to be hoped that this will not happen in the case of Jaguar. The merger with British Leyland did the company damage from which it has not fully recovered, although there are now signs that, under Ford's wing, Jaguar is making the right sort of cars again. Certainly another sensational sports car is sorely needed.

In this book the story of the XK begins with the history of Lyons and his company from the earliest days to the arrival of the XK, concentrating on the main developments and highlighting the arrival of personnel who would become key figures in the development of the XK and its engine. The path towards the XK is shown in the context of the products that went before it. Any story of the Jaguar XK series cannot help but be much biased in volume towards the XK 120 and this is most obvious in the case of competition history.

Jaguar's current owners should remember that Jaguar's success came about by offering outstanding looks, performance and comfort at a price that cars of similar stature could not begin to match. For an illustration of these parameters, the XK series provides a perfect example. When you own an XK, whether it needs work or is already in A1 condition, you can claim ownership of a very beautiful and still fast automobile whose wonderful, sweeping lines clothe a car that was the first to bring the name of Jaguar to the fore throughout the world. With its remarkable collection of speed records, and race and rally victories, it is the genuine all-time classic car.

1 The Early Days

William Lyons was born in 1901, the son of a travelling Irish musician who had decided to settle in Blackpool after meeting the woman he would marry. Their first child, the man who would go on to demonstrate unusual ambition and foresight, coupled with exceptional skills in staff management, negotiating and car styling, was apparently merely 'average' at lessons. Perhaps this should not surprise us too much, for many a person who excelled academically and intellectually in their education has gone on to achieve relatively little, owing to their weakness in the interpersonal skills required in the real world. Lack of intelligence cannot have been the problem for William Lyons – perhaps he had indifferent teachers who were unable to interest him sufficiently.

He became a trainee at Crossley Motors and began to study engineering at Manchester Technical College in the evenings. However, he would not last long at Crossley and soon found himself helping out in his father's piano repair business, before dropping into the retail motor trade. This would all be good training for his eventual path through life and already he was dreaming of becoming his own boss, as his father had been. For some time a keen motor-cyclist, Lyons was particularly intrigued by a newly arrived neighbour who was making stylish looking sidecars in his garage and apparently selling them without difficulty.

A very genial fellow, William Walmsley was some eight years older than Lyons and was assisted in this cottage industry of sidecar building by his wife and three of his sisters. That the first creation was named 'Ot-as-ell' tells us something about Walmsley's jovial nature, but it would not be too long before he was producing a model that he named 'Swallow'. Popular as Walmsley's sidecars were proving to be, it seems as if he had no serious plans to operate on a grander scale. Things were soon to change, however, when the younger William ventured across the road to investigate. He was impressed enough to order a sidecar for himself.

THE SWALLOW SIDECAR COMPANY IS BORN

Sensing the opportunity to develop the business, Lyons put the idea of a partnership to Walmsley, but it did not at first generate much enthusiasm. However, he persisted and with a little support from the older man's now expectant wife, it was not long before Messrs Walmsley and Lyons senior were arranging that now well-known overdraft of £1,000 to help their sons form the Swallow Sidecar Company in 1923. It seems that William Lyons had some reservations about this undertaking, simply because he was slightly worried about his new partner's commitment to entering the world of big business. He had no doubts about his own.

It was not long, however, before they had added coachbuilding skills to their rapidly

The Early Days

William Lyons, 1901–1985.

expanding sidecar business. Indeed, so fast had the business increased in size that the operations were spread over three sites and Walmsley's father had to provide a very necessary improvement in the running of the still fledgling concern, by generously leasing to them a large building that he had bought. The company was now to be known as the Swallow Sidecar and Coachbuilding Company. It was 1926 and they were firmly into second gear.

History, of course, often hinges entirely on actions that were of no apparent historical significance at the time. The decision by the Walmsley family to move to the same street as the Lyons family is certainly one such, and the decision by William Lyons to purchase an Austin Seven soon after he married Greta Brown may be another. These cars were designed to be a scaled-down version of a full-sized car, and they thus enabled the less well-heeled motorist to avoid the notorious cycle-car when taking the next step up from a motorcycle. This was the main reason for the baby Austin's success, just as the Mini would help drivers avoid the equally notorious bubble car some 35 years later.

Many a future motor manufacturer entered this risky profession as a result of dissatisfaction with a product they owned, coupled with a belief that they could do better: Henry Royce with his Decauville and Ferrucio Lamborghini with his Ferrari are two obvious examples. Lyons was another and he was soon designing ways to improve on Herbert Austin's wee product, if only with cosmetic improvements at this stage. In fact, he had been wondering just how much luxury could be added to the austere but appealing little car.

Lyons was not in fact the first to add some hint of luxury to a mundane people's car. MG were putting bodies on Morris cars, and Gordon England were already making their own bodies for the Seven. The recent move to larger premises enabled Lyons to think about offering his versions of vehicles with four wheels. Apart from the more fashionable bodywork, which was an early demonstration that Lyons was exceptionally gifted at putting curves into car shapes, these Austin Swallows of early 1927 also boasted two-tone paintwork with a variety of colour schemes, normally the preserve of far more exotic machinery. More comprehensive instruments were also fitted. Lyons showed his new baby to the retailer P. J. Evans in Birmingham, who promptly asked for fifty. This was very

The Early Days

encouraging, but then a visit to Henlys in London's Great Portland Street immediately yielded a request for five hundred – a staggering number for the small company. Lyons accepted this order like a flash, although he subsequently admitted that he was desperately worried how they would manage the target of twenty cars per week. On hearing of this order, his less ambitious partner insisted he had gone mad. Nevertheless, the enforced expansion of the firm took place and the delivery target was met.

Lyons now had the bit between his teeth and several other Swallow bodies would find their way on to other makers' chassis. It is sometimes reported that his wrath at what he perceived to be Herbert Austin's discourtesy during a meeting caused him to look beyond Austin-supplied chassis. Swift, Morris, Wolseley (their first association with a six-cylinder), Fiat and – significantly – Standard would all have their turn at being adorned by a Lyons-style body. Clyno and Alvis chassis received the Swallow treatment but seem not to have progressed past the prototype stage. In almost every case the looks were an improvement over the manufacturer's original version of the car.

Some years later just about every attempt by other coachbuilders and styling houses to improve on the original Lyons product failed quite miserably. Apparently Lyons produced a lot of rather ugly creations on the way to signing off the final

The first two-seater, a 1927 Austin Seven Swallow.

The Early Days

A non-Lyons attempt at styling a Jaguar – an XK 140. On the reverse of this photograph someone has written '... designed by Raymond Loewy (after a nightmare?)'.

shape as he wanted it, but it would not be easy to overrate him as a car stylist and the Jaguars that have appeared since he ceased to have overall styling control have been very disappointing.

ON THE MOVE

The success thus far of the Swallow company was causing increasing problems with recruiting sufficient skilled labour. Accordingly, Lyons started examining suitable sites in the industrial heartland of the country, the west Midlands. A disused munitions factory eventually fitted the bill and the firm moved from Blackpool to the town of Foleshill, situated to the north-east of Coventry city centre. This move occurred in 1928 and the firm would not move again for some 23 years.

Soon after this move the association with Standard began, when one of the sporting Swallow bodies was seen on a Standard Nine chassis. This was followed in early 1931 by a larger car altogether, based on a Standard Ensign Six 15hp (sometimes 16hp) chassis. Fifty-six of these cars were made and it would be the largest Swallow saloon to be offered. This use of a Standard straight-six engine was certainly a portent of things to come.

Despite the quite lavish praise that his products were now generating, Lyons was searching for a yet longer and lower look. No current chassis was available to provide

The Early Days

Lyons with the silhouette he required, so he approached the Standard Company through Reginald Maudslay and John Black for a bespoke chassis and engine. Broadly speaking, this would be a Standard Sixteen chassis with modifications to the springs and hangers. Appearing at the 1931 Motor Show, the resulting car was named SS for the first time. This new six-cylinder 16hp became the SS 1 (also offered with a 20hp engine) and the altogether smaller four-cylinder 9hp car, the SS 2.

Much has been written about what the initials 'SS' stood for, the debate given mystique as the factory (probably deliberately) never explained their meaning. The main company name had, after all, been 'Swallow Sidecar' and it was not the best selling point to have the name 'Sidecar' when trying to sell a proper car, let alone one with sophisticated intentions. So the company had quietly dropped 'Sidecar' from its title at the end of 1929, becoming the Swallow Coachbuilding Company. Since these new cars were really 'Swallow Sports' and since they were also 'Standard Swallows', 'SS' neatly fitted both titles. (There had been Brough Superior motor-cycles that carried Swallow sidecars as well the designations SS 80 and SS 90, so these initials would certainly have had some cachet to Lyons and Walmsley, both once very keen motor-cyclists.)

The first SS 1 looked rather faster than it was, although it would be unkind to call it a sheep in wolf's clothing. Also, its proportions were not quite right, the bonnet being a little too long and the windscreen a shade too vertical. It was known that Lyons was not at all happy about the styling, which had been done while he was convalescing. On seeing it for the first time he is reported to have said that the passenger compartment looked like a conning tower, but he had to give in because of the imminent deadline of the Motor Show. It would thus be fair to say that the SS 1 was hurried in its execution, a claim later borne out by the large number of niggling problems in service.

Nevertheless, the SS 1 made the front page of the *Daily Express* – to Lyons's astonishment – and despite the car's shortcomings, it could certainly be described as a sales success. One year later, in the autumn of 1933, a revised 1934 model SS 1 appeared with several mechanical improvements,

Long and low – the beginning of the 'Lyons line'. A 1934 SS 1, the 'Mark II' version.

more rear leg room and rather more attractive styling, Lyons having been able to provide his input this time. This car would start to generate significant sales, a total of more than 4,750 examples being sold in the next three years. At the end of 1934, SS Cars was floated as a public company, raising £85,000 on the share issue.

Another event of significance took place at the end of 1934, when William Walmsley left the company. It had often been noted that he found himself almost reluctantly swept along by his younger partner's dynamism and drive. This is no criticism at all, for some people like, while travelling through life, 'to have time to sniff the flowers along the way'. He was only 42 and would spend time in various ventures (such as caravan manufacture) and his hobbies. When he died in 1961, his former partner did not attend the funeral.

THE SS 90

In the spring of 1935 there appeared the first true high-performance two-seater offered by the firm, the SS 90. Some early Austin Swallows were two-seaters, as was the six-cylinder Wolseley Swallow Hornet, but this car was Lyons's real foray into the high-performance arena. It had the usual 2½ litre (2,663cc) Standard side valve engine, with slightly raised compression but still rated at 20hp.

It possessed several features that were rapidly becoming the factory's trademarks and that would be seen on the next two two-seaters, the XK and the E-type, from the Coventry company, the only two all-new Jaguar sports cars in sixty years. The SS 90 had a very long and low sporty appearance – the sort that suggests great speed when at rest. It was remarkable value for money and was propelled by a straight-six engine with a stroke of 106mm and seven main bearings, which helped to give that characteristic smoothness. Although fast, it was not fast enough for Lyons's taste as the performance leader of the range. Fewer than two dozen of the SS 90 model were made, such was the speed of the developments at the Foleshill works.

To improve performance in the top of the range cars as a whole, a straight-eight and even a supercharger were considered and rejected. This latter idea would have to wait until 1994 before first appearing on a factory product, where it enabled the 3,850lb (1,750kg) XJR four-door saloon to out-accelerate the 2,650lb (1,200kg) 3.8 E-type. It was then decided to commission Harry Weslake to design an overhead valve (OHV) cylinder head to fit straight on to the original Standard block.

Acknowledged as a leader in the field of cylinder head design, Weslake produced a 'cross-flow' layout, where the inlet and exhaust ports lay on opposite sides of the head. This was quite advanced for its day – Ford were very pleased about the introduction of such heads more than thirty years later and used this feature strongly in their advertising. Weslake had already been called in on a consultancy basis by Lyons to help solve a problem with the new aluminium heads, unreliability having manifested itself during the abnormal (by road standards) stresses experienced by SS 1 engines in the Alpine Trial. This attack of unreliability had been brought about by last-minute modifications to increase performance, modifications that were not properly tested before the competition. (Alas, this makes the 1952 Le Mans debacle even sadder.) At least a team award would be theirs in 1934, this despite more gasket problems.

Weslake was well under way with the OHV project when Lyons appointed a man

The Early Days

who would become the most significant staff member, after himself and his former partner. Bill Heynes was appointed to head the newly formed Engineering Department, having come from Humber. It was the task of this new recruit to help bring the new car range into full production. Considering the short time-scale with which he had to play, he was truly in at the deep end.

These new SS cars were to carry the suffix 'Jaguar' for the first time: Lyons had employed a London advertising agency to present for his approval some names that would echo the desired blend of grace and pace. He had to seek permission from Armstrong Siddeley as they had used this name on an aero engine. It was granted readily – ironically it would be Jaguars, more than any other cars, that would eventually put Armstrong out of business.

THE SS 100

The new OHV engine managed to produce around 100bhp, which was about 25bhp more than the side valve unit of the same capacity, although it would still be rated at 20hp for taxation purposes, thanks to the rather antiquated RAC formula for calculating horsepower (*see* panel). As well as gracing the big saloons, this new engine also found its way into a slightly altered SS 90 two-seater, now designated the SS 100.

Announced only six months after the 90, the SS 100 also had improved brakes to help with the increased urge from its new OHV power unit. These were now Girling rod-operated, instead of Bendix cable. This gave a more positive feel as the rods avoided most of the elasticity in tension that arose with cables. Appearance was improved, too, with alterations to the rear of the car and subtle changes to the front.

RAC Horsepower Rating

Motoring journals in the 1930s never seem to have given the actual power outputs, they merely quoted RAC-rated horsepower, a rather inaccurate measure. RAC horsepower rating is not directly related to the actual power output of an engine, but is calculated using bore size, number of cylinders and a constant. For example:

$$\text{RAC hp} = \frac{d^2 n}{K}$$

Where:
d = diameter of cylinder (mm)
n = number of cylinders
K = 1613

The 100 was certainly a beautiful car now. The essential point about the 100 was that the new engine gave an SS product the performance to match its looks. Its top speed was just short of the magic three figures. When the engine grew to 3½ litres in September 1937, the SS Jaguar 100 was, with its genuine 100mph performance, just about the fastest standard volume-production car one could buy. In this respect, it began a trend followed by the first XK and the first E-type.

As far as competition history goes, this new SS Jaguar open two-seater sports car started another tradition by winning a major event at its first attempt, just as the XK and the E-type did some 13 and 25 years later. This was the 1936 Alpine Trial, where it was driven by Tommy and 'Bill' Wisdom, the formidable husband and wife team. The SS 100 was more at home in rallies and although it did win several circuit races, it did not manage the quite remarkable number of track successes of the XK 120, nor even as many as the less successful E-type.

The Early Days

An SS 100 owned by Mrs V. M. Hetherington, taking part in the Scottish Rally, seen here refuelling at Inveraray in the late 1930s. The makings of the XK wing line are already in evidence.

Incidentally, on the few occasions when William Lyons himself took part in a race, he always seemed to be the fastest driver on the day. Obviously, his exceptional ambition did not confine itself solely to manufacturing. His turn of speed recalls the exploits behind the wheel of two other car bosses – Frank Williams of the current Grand Prix team and the late Colin Chapman of Lotus. Both were quite likely to run away with any celebrity race in which they took part. All part of the desire to succeed, no doubt.

In 1937 there was a significant alteration in the manufacturing process: a change to steel bodies for the saloons. This caused a backlog of orders until the new system was able to operate smoothly, causing much lost production in 1938. No doubt Heynes found his stress levels rising once again.

It was in August 1938 that Wally Hassan was appointed as a senior engineer to assist the beleaguered Heynes. Hassan's credentials included working for Bentley, ERA and Thompson and Taylor, where he was involved in Reid Railton's land speed record car, successfully piloted by John Cobb. He would take up the post of Chief Experimental Engineer, his main task being to start work on a future independent front suspension. The stage was already being set for the XK.

With successes in competition, coupled with an unrivalled combination of looks, speed and value, by 1939 the SS Jaguar

The Early Days

range had become highly desirable, and not just in Britain. This was a record year, with total home sales reaching 84 per cent of the 1993 levels. Had the war not started in September, these sales figures would surely have gone on to exceed those of more than fifty years later. It is startling to observe the rapidity with which Lyons and his company had attained this position: just seven years previously they were still putting bodies on Austin Sevens.

This car maker had charged irresistibly through two depressions without apparently breaking step, while all around others were struggling to stay afloat and many more simply slid under for ever. However, the economic conditions may have helped the young company in a couple of ways. First, many people in the financial squeeze traded down (against the trend) to the Austin Swallow, to avoid the extra expense of a larger car while retaining a pretence of luxury and some individuality. Second, in times of high unemployment skilled labour was easier to find by those firms that were still managing to expand. Nevertheless, to have continued on an upward trend throughout the 1920s and 1930s is indeed a forceful demonstration of the inherent strengths of Lyons's products and business strategy.

As the dark shroud of war wrapped itself around the country, choking off car production, William Lyons still had a geat deal to be proud of. He was thirty-eight years old and employed more than one thousand staff, but his best was yet to come and five years of war merely checked this extraordinary progress for a while. After the end of hostilities it would not be long before Jaguar products would become truly world-beating.

2 Developing the XK Engine

The Foleshill factory again found itself a centre for war production, for the works were now commissioned to build parts for aircraft such as the Mosquito, Stirling, Whitley, Spitfire, Lancaster and even the Meteor jet. The factory also became the main suppliers for the service and repair of the Whitley. Parts were also made for the Cheetah engine, a radial of seven cylinders. Ironically, SS Cars Ltd would also make around 10,000 of various types of sidecar for the war effort. The most common unit to be made was an amphibious trailer, of which more than 30,000 were made.

One significant benefit of all this war work was the more up-to-date equipment sent in by the War Ministry. Another was the re-hiring of Wally Hassan, who had felt under-utilized at the start of the war and had moved to a post in Bristol. Back in Coventry he would work on a small parachutable self-propelled vehicle for the War Department, alongside Claude Baily (formerly of Anzani and Morris) who had joined the factory at the start of the war as chief designer to Heynes. The teaming up of these two on war work would be of great significance to the development of Jaguar's first all-new engine. Their little War Department vehicle would not go into production, as parachute design had improved enough to allow a Jeep to be dropped from an aircraft and remain assembled. Nevertheless, Baily, Hassan and Heynes acquired some invaluable knowledge because their 'mini Jeep' was of unitary construction, pre-dating Jaguar's first official attempt by a dozen years or so. (Furthermore, its rear suspension had more than a little in common with a certain sports car that was unveiled at Geneva in 1961.)

FIRE-WATCHING DUTIES

Heynes, Hassan and Baily, encouraged by Lyons, were mulling over what sort of engine their company would build when normality eventually returned to the world. It is always stated that the main parameters of the engine were settled while they were on fire-watching duties towards the end of the war. Apparently Lyons had arranged for all four of them to have coincidental fire-watching shifts on a Sunday evening, for this specific purpose. They had probably also given the new motor plenty of thought earlier, while working on War Department contracts in the Development Department. An entirely in-house engine would enable them to become fully independent car makers – an ongoing aim for Lyons.

One goal was that the engine should be powerful enough to push a full-sized saloon through the air at 100mph (160km/h) and

Developing the XK Engine

yet be tractable, even docile. Lyons had suggested that the new engine should match the peak achieved from the 3½ litre Standard-based pushrod unit. Installed in the fastest pre-war SS 100, over 160bhp had been obtained, running on a brew that amounted to a sort of liquid Semtex, and the new engine's task was to match that output in standard tune, with plenty more to come from future development. Lyons also wanted the engine to look glamorous. (He has been cited as the first person to style an engine's appearance, but whoever wrote that had not lifted the bonnet of a Bugatti.)

Various layouts were considered, of four, straight-six and V-eight configuration. The four-cylinder was chosen to power the smaller of the two distinct sizes of saloon car, which had been chosen to be the main basis of the range. Testing of prototype units would have to wait until the war was over, however.

Lyons had another good piece of fortune nearing the end of the war when Sir John Black told him that it no longer fitted into Standard's plans to supply him with the OHV engines on an exclusive basis. Lyons immediately arranged with Black to buy the machine tooling and very quickly had it all delivered safely to Foleshill. It is reported that Black wanted to change his mind soon after and was – of course – rebuffed. An offer by him to form a separate company with Lyons was also turned down. Black was rather incensed at this, considering

What Lyons was after: an engine with an appearance to match its performance. This is from an early aluminium-bodied car. Strangely, the cylinder head is painted matt black.

the co-operation that had hitherto existed between the two companies, and threatened to buy the ailing Triumph concern with a view to generating enough competition to force Lyons out of business. Lyons was unmoved, having been unimpressed with its prospects when offered that company himself in 1939 when the Official Receiver had been stalking it. Black managed to carry out the first part of his threat, but not, of course, the second. Considering the products made by Triumph since the war, the TR2 excepted, Lyons's foresight must again be congratulated.

In February 1945 it was announced that the company title would henceforth be Jaguar Cars Ltd. The notoriety that the initials 'SS' had earned before and during the war meant that a hasty dropping of these initials became necessary. Nevertheless, they would reappear in 1957 on the handful of XKSS models that were made, based on the D-type. Although the XK 120 was catalogued as the 'Super Sports', this title was never abbreviated to its initials.

POST-WAR PRODUCTION

With development of new cars (and a world-beating engine) under way, Lyons was shrewd enough to bide his time and continue to offer pre-war models, which began to come off the lines in October 1945.

No drop-heads or two-seaters were offered at first. Lyons was confident enough not to introduce the new models until they were fully ready – a concept that some other British car makers would not understand in the ensuing years.

It was in 1946 that another significant appointment was made to the company. He was a friend of Wally Hassan and had the highest credentials in the Service Department and as a racing mechanic. He would become another of those persons intrinsic to the success of the XK series and Jaguar Cars generally. The key personnel of Lyons, Heynes, Weslake (as consultant), Hassan and Baily were now joined in February 1946 by Raymond 'Lofty' England.

Appointed as Service Manager, Lofty England had been apprenticed to the Daimler service department in north-west London for five years and had then gone on to become Assistant Service Manager for Alvis. His experience of motor racing, like Hassan's, was considerable. He had competed a few times, not without success – coming second, for example, in the 1932 RAC Rally in a V12 Daimler. A list of those with whom he had been involved in racing car preparation reads like a Who's Who of pre-war motor racing: Sir Henry Birkin, Dorothy Paget, Lord Howe, Whitney Straight, Giulio Ramponi, Dick Seaman and Prince Birabongse of Siam, who raced under the name of 'B. Bira'. Lofty England had credentials!

The inflationary cost of the war is strikingly demonstrated by a comparison between the immediate pre- and post-war prices for the same models.

Models	1939 price	1946 price
1½ Saloon	£298	£684
2½ Saloon	£395	£889
3½ Saloon	£445	£991

Developing the XK Engine

*'Suits you, Sir. Just get your father to sign the necessary and this XK 120 will be yours.'
Super salesman meets royal punter.*

EXPORT DRIVE

Another piece of fortune for Lyons – although he may not immediately have seen it as such – was the compulsory export quota in operation at that time, thanks to the austere plans of the post-war government. 'Export or Die' was one of the current slogans and the quota was a staggering 75 per cent as Britain's economy struggled to recover from six years of industrial turmoil. Tied in with this was the method of steel allocation. Strictly rationed for home consumption, the allocations were more generous if the steel-made goods were to be exported. Although not a great exporter before the War, Lyons lost no time in drawing up extensive export plans, which were then presented to the government. The all-important steel allocation was made.

Offering left-hand drive for the first time, Lyons himself went hunting sales across the Atlantic and was very successful: it would not be long before a huge rise in sales would occur. The average yearly export figure for the five years prior to the war was 216 cars, with a peak of 252 in 1939. This represented a mere 8 per cent of total production. In the five years from 1946, the yearly export average was 1,718 cars, a more than eight-fold increase in numbers and a growth to an average of 44 per cent of production.

Good fortune is often balanced with bad and 1947 was certainly a difficult one for Jaguar. At the beginning of the year there was a fire in the stores that did an estimated £100,000 worth of damage. This equated to the purchase price of about one hundred of the company's most expensive models. The rest of the year saw production struggle on through coal shortages and power cuts, as well as the crippling shortages of steel. Despite all these difficulties, production continued to rise.

During 1948, Standard ceased making the four-cylinder, 1,776cc engine for their own use and this caused Jaguar to drop the car powered by this engine. Curiously named the 1½ litre, it had been selling better than it should have been, thanks to the petrol rationing in force. It was not offered in America, and Continental markets took almost no notice of it at all after the war. Production of this smallest car in the Jaguar range stopped at the beginning of 1949.

DESIGNING THE NEW ENGINE

Not long after wartime production ceased, the factory began to bring its new power unit into reality. Valve operation by twin overhead camshafts was decided upon after various layouts were discussed and some actually made. One goal was that peak power should be at 5,000rpm – quite high for a large engine at that time, and still proving quite high today for an engine with such a long stroke. Such piston speeds and rotational velocities would need first-class oil delivery and an engine under test was run at 5,000rpm for 24 hours under full load. This would use as much as 500 gallons of petrol – a costly amount for one day's consumption. Indeed, various four-cylinder prototypes were assembled, starting with a twin-cam, given the code name XF. The XA to XE units apparently did not make it off the drawing board.

Jaguar would not be the first manufacturer to offer twin overhead camshafts on a series production road car, for this honour falls to a 1922 Salmson. A small coincidence is that the model so propelled was named the D-type.

The XF was a mere 1,360cc with a bore

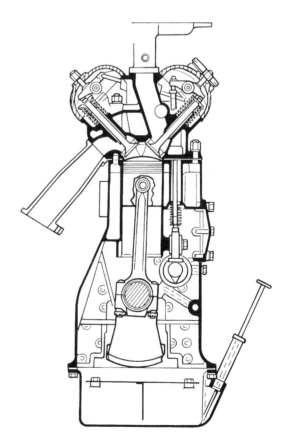

Rather different: the XG, based on the BMW pushrod design. The single camshaft can be seen just to the right of the big end journal.

Developing the XK Engine

of 66.5mm and a stroke of 98mm. It was soon discovered to be generally lacking in strength, mainly in its bottom end. The XG was also a four-cylinder engine, based on the 1,776cc Standard unit, with dimensions of 73mm bore and a prescient 106mm for the stroke. It was cast with the BMW system of cross-pushrods, giving the then state-of-the-art hemispherical combustion chambers of the twin overhead camshaft design, but using only one camshaft. The valve gear was deemed too noisy, although the hemispherical combustion chamber layout was now a certainty.

Next was the accidentally prophetic XJ engine, which was reckoned by Heynes to be the true parent of the XK unit. This had grown larger still to 1,996cc (80.5mm x 98mm) and making a six-cylinder version of this – naturally with a seven-bearing crankshaft – with an 83mm bore and the same 98mm stroke gave 3,186cc. This was a far smoother power unit than any of the previous 'fours'.

The unit now arrived at was intended to replace both the previous 2½ and 3½ six-cylinder units, but it was a little lacking in good low-speed torque, already recognized as a Jaguar forte. Indeed, it proved to have less twisting power at the slower revolutions than the old 3½ litre engine, which delivered its maximum torque at an impressively low 2,300rpm. In stretching the crankshaft throw to 106mm, Jaguar ushered in the arrival of some now famous dimensions. This unit, with its twin overhead camshafts (OHCs) and hemi-head design with its new swept volume of 3,442cc, was given the code name XK. A smaller four-cylinder version of the engine would be offered, to cater mainly for the financially squeezed home market of postwar Britain.

The hemispherical head's main advantage is that it permits just about the best

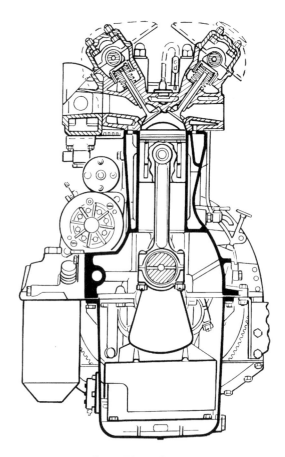

XK engine, end-on. The coil is seen immediately above the starter motor on the left, making it look like an improbably early pre-engaged type.

flow through the valve throat, which – other factors being correct – will enable the engine to reach higher revolutions and so develop more power. Of only slightly lesser importance, turbulence can be easily controlled by port shape, and a degree of inlet swirl will, for example, give very good combustion of the charge, especially with a central spark plug with a short flame path. These factors provide the ability to employ a weaker mixture yielding good fuel economy (relative to the early 1950s). Another advantage of well controlled combustion is the absence of running-on or self-ignition. The exhaust valves can also be cooled relatively easily, thanks to a good flow of cooling water that can be had around the seat, and manufacture of this type of chamber is very simple, the correct hemispherical shape being obtainable in machining by one movement of a single cutter.

This finalized design for the cylinder head was favoured over several others and was the state-of-the-art in combustion shape at that time – outside Grand Prix machinery, that is. If Jaguar's post-war fame rested very firmly on its XK engine, the success of that engine was in its cylinder head, and the success of that head had much to do with Harry Weslake. Nevertheless, Heynes was still Jaguar's Chief Engineer and those who attended his reading of his paper on the XK engine to the Automobile Division of the Institute of Mechanical Engineers on 14 April 1953 heard about the various design problems and general development of the XK engine straight from the horse's mouth. The extract shown in the panel overleaf is reproduced by kind permission of *Autocar*, and it shows that a very considerable quantity of experience, scientific knowledge, technology and good old-fashioned thinking went into the XK engine. With its aluminium cylinder head of highly advanced design, and the very substantial long-stroke bottom end running in the

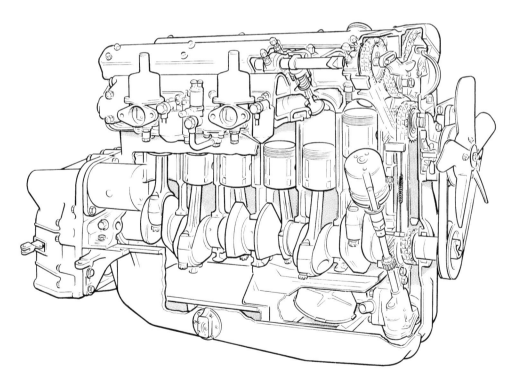

XK engine. Note the bottom timing chain tensioner. This would soon be changed from the rather feeble spring to an hydraulic device. Also note the absence of a crankshaft damper.

The Development of the XK Engine

Regarding cylinder head material, the high conductivity of aluminium is not its only advantage; another important one is the saving in weight. The bare XK cylinder head weighs 50lb, but in cast iron it would weigh about 120lb. Ease of machining is another advantage, so is ease of handling without lifting tackle. After experiments with high silicon alloy the less costly DTD 424 was adopted, with valve seats of high nickel austenitic cast iron (Brimol). The seats are inserted with the head at 232 deg C and tappet guides are fitted at the same time. Valve guides are fitted as a separate operation with the head at 80 deg C. Sparking plug inserts were fitted in only a few heads in the early stages, but there is no record of a single case where rebushing has been required.

Valve port design contributes much to the high output per litre, and was carried out by Weslake and Co. The basic principles of the curved port with a venturi orifice were established by design and are subject to a patent, but the final shape of the port was obtained by a flow test. Full-scale models of either wood or aluminium are cleaned out or filled up in the port orifice until the maximum flow has been obtained. Experience shows that a larger valve or a higher lift will not invariably produce a higher flow. Very small changes in port shape can produce a large difference in flow, so for accuracy male cores are taken from the model and from these the core boxes are made.

The method for arriving at port shape by flowing instead of repeated bench tests gives a great economy in development time, and, although a certain standard of perfection can be produced by design alone, the author has yet to see a designed head that will not yield a further 10 per cent bhp at the same peak revolutions by the intelligent use of this technique.

Of fairly conventional seven-bearing design, the crankshaft is notable for the large diameter of the main bearings, 2¾in. The shaft is partially counterweighted, the centre bearing being the one given the most attention, other weights being disposed to retain dynamic and static balance of the crank as a unit. It was realized that the torsional vibration period could be a more formidable problem than bearing loads, and that the extra weight of a crank fully counterweighted for each journal would increase this problem.

Particular care is taken in balancing the crankshaft assembly. The separate shaft is first dynamically balanced on an Avery machine. The flywheel, which has previously been balanced on a Micropoise static balancer, is then fitted and the assembly rechecked statically and modified if necessary. A final recheck is taken again after the clutch has been bolted in position.

Torsional vibration is probably the cause of more engine failures, either directly or through its effect on other units, than any other single factor in competitive events. The Metalastic damper is a steel plate to which is bonded, through a thick rubber disc, a malleable iron floating weight, and variations of the weight, the rubber volume, and the mix give a very wide field over which the dampers can operate. At the early stage of development when the engine reaches a test bed, a primary check is made without a damper. According to results sample dampers are fitted and from further test results the final specification is decided.

The crankshaft is of EN16 steel, heat-treated before machining to give a Brinell figure of 270–295. Bearings are thin-walled steel, babbit-lined, but in the 120C engine indium-coated lead-bronze is used, with increased clearance provided by using a thinner bearing.

Journals and pins are ground and hand-lapped to ensure a smooth surface and a finish of about 10 micro-in. Where the oil hole breaks into the pin the edge is stoned away to remove the sharpness and give an oval countersink as a lead-in for the oil. The feed-holes are drilled through the pins at 90 deg to the throw, and the pins are drilled diagonally to form a sludge trap, the holes being sealed by taper threaded grub screws, additionally held by centre punching the crank webs.

The simplicity of the cylinder block, the minimum of machining it requires, and its comparatively light weight are valuable machine shop assets on this heaviest portion of the engine. The bores are integral and the wear figures are extraordinarily good, mainly by reason of the freedom of the bores from extraneous stresses and uneven temperatures, but also the use of a chromium-plated top piston ring, which tests have shown to reduce wear by at least 50 per cent. The crankcase is split on the crankshaft centre line, the author holding the opinion that the extra stiffness of the flange at this position is valuable. A feature is the tie between the main bearings and cylinder head, webs from the head studs being carried down inside the water jacket to join the webs carrying the main bearings.

To lubrication the author devoted considerable attention, for it is one of the most important features for continuous high-speed operation. A normal gear type pump is used, and the relief valve was normally in the pressure filter but was moved to the pump and recirculates the oil. This reduces the quantity of oil passing through the suction pipe and the tendency to cavitate at the pick-up. It reduces the speed of the oil passing through the delivery pipe and pressure filter and avoids an excess of oil from the by-pass lying on the top of the baffle. This is similar to over-filling the sump, which can cause a loss of over 20 bhp at speeds of 5,000 rpm. The oil delivered at high speeds is considerably in excess of the needs of the engine and unless some such control is provided such a state must inevitably exist.

All oil passages are as large as possible to reduce the speed of flow and to ensure that all bearings are equally supplied. Where possible the corners of holes are broken by counter-sinking, as the author believes that much of the cast iron which passes into the bearings results from the sharp, brittle edges being scrubbed away by the fluid. Each main bearing is separately fixed and oil passes through a drilling in the shaft to the pins, except that in view of its higher load the centre main bearing does not feed the big-ends, these being supplied from the intermediate, and front and rear journals.

The feed to the big-end is through a hole drilled at right angles to the web centre line, because it was found that the original drilling, feeding the rod at the largest radius of the crank, was acting as a very efficient centrifugal pump. Accordingly at speeds above 5,000 rpm there was a tendency to lose oil pressure. The repositioning of the feed hole reduced requirements at 5,000 rpm by half.

A factor not always appreciated with aluminium alloy rods is the adverse effect on oil pressure control owing to the high coefficient of expansion. The increase in diameter of the big-end eye when hot practically doubles the clearance and the loss from this point. This was one reason for reverting to steel rods.

After the hemispherical combustion chamber was decided upon at least a dozen layouts for valve operation were considered before settling on the simplest system, twin overhead camshafts with cams operating directly on the tappets. The author outlined the advantages as low reciprocating weight, permitting valve spring strength to be reduced, absence of wearing surfaces between tappet and valve, elimination of rocker side thrust and, therefore, valve stem and guide wear, and protection against excessive oil consumption given by the inverted tappet. The camshaft drive on production engines closely resembles that used originally. A simplified layout with a single chain was tried, but although satisfactory in operation it produced a high-pitched whine which defied cure.

Camshaft sprockets have internal serrations which, in combination with the 22 chain teeth give a vernier adjustment for timing. The sprockets can be removed and held on dummy spindles with the chain *in situ* so that the timing is not lost when removing the cylinder head.

The cooling system is pressurized to 7 lb per sq in by a spring-loaded filler cap. A centrifugal pump forces the water through a gallery pipe which runs the full length of the block on the exhaust side. From there it passes into the head below each exhaust port, flows round the exhaust valve seating across the head, past the inlet port, and enters the jacket of the water-heated induction pipe and so past the thermostat to the radiator. There is no block circulation from the pump but convection is permitted through large orifices into the head on the inlet side.

An unusual cooling problem arose by reason of the wide speed range of the engine. The pump runs at 0.9 engine speed, and at 6,000 rpm a severe tendency to cavitation was experienced. After much research the impeller was modified, the blades being given an improved angle of entry and cleared back at the centre to prevent possible rejection of the water by the unshrouded part of the blade. When cavitation is bad steam is formed in the eye of the pump and flow stops completely, with very dire results. The effect of water temperature on both flow and cavitation is not always recognized; a few degrees' rise in ambient temperature can completely upset a system that is near the margin.

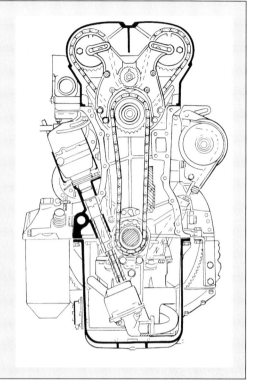

XK engine, showing chain drive to camshafts and the way the distributor and oil pump share the same drive, taken from the nose of the crankshaft. The ghosted item is the fan belt.

chrome-iron block, the engine has been described as 'a racing top end, with a lorry bottom end' – a little bit dismissive, but not entirely inaccurate. Bottom end failure, even under racing conditions, would turn out to be a very rare event for a properly maintained XK unit: a tribute in part to the great reserves of strength in the EN16 steel crank. (EN stood for 'Emergency Number', a wartime grading system.)

FLYING IN BELGIUM

Meanwhile, one of the experimental four-cylinder engines was making the news. In the late summer of 1948, something of a sensation was created when Lieutenant-Colonel A. T. 'Goldie' Gardner flew down the Jabbeke to Aalter main road in Belgium – now part of the Ostend to Brussels highway – at 177mph (284km/h). This he did in his MG-based special, EX135 (found by *Autocar* in May 1982 to have a drag coefficient of 0.26), propelled by an XJ version of the new Jaguar twin-cam. Strictly speaking, it was not one of the test units but a development of one with a compression ratio of 12:1. This was heady stuff when the contemporary 3½ litre Jaguar pushrod unit had a 6.75:1 ratio and the first XK unit was introduced at 7:1.

Rumours of the new Jaguar models now shifted up a gear or two and on 1 October 1948 the Mk V was officially announced. This would be the first of Lyons's products to sport all-hydraulic brakes and – more importantly – independent front suspension, incorporated into its all-new chassis. It was a very handsome car indeed, with its

Developing the XK Engine

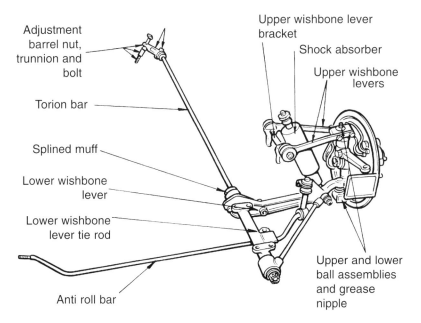

The first-class front suspension, as fitted to an XK 120. It would remain essentially unaltered throughout the XK series.

Jaguar styling ancestry still very intact.

Hassan's front suspension comprised two wishbones that held a large-diameter ball and socket joint top and bottom, which in turn held the hub carrier. The use of ball joints at both ends of the hub carrier meant reduced conflict in the suspension geometry while it moved through its operating arc from full bump to rebound. The springing medium was by torsion bar, attached to the substantial part of the chassis at the bulkhead, where its loads could be more easily absorbed, thus avoiding most of the twisting effects around the front of the chassis where it was thinner. The other end of the torsion bar was splined to the bottom wishbone at its inner end.

Torsion bars are often spoken of with a certain exaggerated respect, perhaps because they have not always been the most common form of springing. A torsion bar is merely a straightened-out coil spring and works in the same way. (It would be more accurate to label a coil spring a curled-up torsion bar.) However you look at it, Hassan's independent front suspension was a vast improvement on the cart-springing of all the previous Jaguar and SS cars. It was the main ingredient that endowed this new car with a ride comfort that many motoring journalists claimed set new standards.

When introducing this new range of Mk V models, in both 2½ and 3½ litre forms, to the motor trade the night before the official announcement, Lyons made references to a new sports car. But since there was still no sign of the twin-cam that was glimpsed the previous month in the Gardner record-breaker, and as it was thirteen years since an all-new Jaguar two-seater appeared, rumours were now rife. Nobody expected to see it quite so soon, however. Indeed, Lyons had deliberately introduced the Mk V before the Motor Show to avoid detracting from the impact that his new sports car might make. But could the Mk V really have stolen the XK's thunder?

3 The XK Leaps In

The 1948 London Motor Show at Earls Court opened to the public on 27 October. A low, long, streamlined bronze XK 120 open two-seater shared the Jaguar stand with the new Mk V, a six-cylinder XK engine and a four-cylinder version, purportedly to power the XK 100. (It was actually the engine taken from the Gardner record car.) With its styling redolent of, and improved upon, the BMW placed third in the 1940 Mille Miglia, the new XK created more of a sensation at a British motor show than any other sports car, ever.

Had the 1961 E-type been unveiled in Britain, such a claim might be put less boldly. As to which of these two deservedly legendary Jaguar sports cars created more of a stir at their first appearance at a motor show, it is hard to decide. Perhaps it was a dead heat. Just as with the E-type, the excitement for the XK was for three inter-related reasons: its performance, its beauty and its price.

It was the 120mph (195km/h) top speed claim for the XK 120 that caused such a sensation – and a certain amount of scepticism. A survey of the maximum velocities of bread-and-butter saloon cars that were currently available at the time of the XK's introduction makes an interesting comparison with the two-miles-a-minute boast of the larger-engined XK on offer. Taking an upper limit of 1,500cc engine size, the average top speed for four-door saloons ('family' cars) being produced in October 1948 was 62mph (100km/h). Perhaps little further explanation is needed of the thrill that was generated at the time by the speed claimed for the XK 120.

To see how other sports cars performed at the time is perhaps more surprising. According to figures in *The Autocar*, the 1948 Lea Francis Sports two-seater cost £10 more than the Jaguar and had a top speed of 87mph (140km/h), while the similarly priced Allard two-seater with its 3.6 litre V8 may have had a larger engine than the XK but only managed 86mph (138km/h) on test. Both cars had acceleration that also would turn out to be considerably inferior to the XK 120. It is not surprising that people flatly did not want to believe Jaguar's performance claim for their new 'Super Sports' model.

NOT JUST SKIN DEEP?

As for its looks, the car's appearance was greeted with unrestrained enthusiasm. It was regarded as a beautiful machine on its introduction, but approaching fifty years on its visual appeal seems, if anything, even greater. Aluminium was used for all the external bodywork with steel for the main structures underneath, such as the scuttle area. The aluminium panels were supported by a laminated ash frame. On later cars, introduced in April 1950, only the bonnet, doors and boot lid would be aluminium.

The overall shape is long and low, having that delightful downward fall from the front wings and the more abrupt rise as the sweep crests the rear wheels, to form the car's rear haunches. It is this aspect of the XK 120 that copies, perhaps accidentally, the wing line of the aforementioned BMW. The front of the XK is far prettier, however, than that of the one-off Bavarian machine.

Viewed directly from behind, but from a little higher than normal vision, the feline grace of these rear haunches is striking. The overall shape has not been compromised by any unnecessary adornment, having the most slender of bumpers, a delicate windscreen (reminiscent of a 1935 Auburn 851 Speedster), no external door handles and covered-in rear wheels. Quite simply, time has revealed it to be a classic, one of those rare shapes that never really went out of fashion completely. The three-quarter front view of the open two-seater shows it to be a masterpiece. Considering the haste with which it was styled, the car's beauty may be considered even more remarkable.

December 1948 advertisement from Motor Sport.

Sleek and speedy. The Motor Sport *road test car.*

The XK Leaps In

True beauty needs no adornment.

As for its cost, the basic price was £988, as shown in Jaguar's own advertisement at the car's introduction, headed 'Sleek Beauty and Speed'. The price with purchase tax was £1,263 and this was extraordinarily low, given the car's performance. This price was only four times that of the cheapest contemporary Ford. At the other end of the scale, the XK 120 was only 23 per cent of the basic price of the cheapest Rolls-Royce.

Yet another way to assess the remarkable value of the 1948 XK (both the 100 and the 120 were marketed at the same price), would be to look at the cost of its competitors. The car that comes closest to fitting the description of a competitor seems to be the Mille Miglia version of the BMW 328. Termed a Frazer Nash BMW for sale in Britain, this car cost more than twice the XK, being listed at £2,000 basic. Many other sports, or just sporting, cars of inferior performance still cost more than the Jaguar.

If you put together all that Jaguar were offering in the XK 120 and attempt to translate it into the mid-1990s, then they would be offering a completely untemperamental sports car that sets new standards of ride comfort for its type, clothed in a superbly attractive body and that reaches 200mph (320km/h) – faster than any other standard production car. It would cost somewhere between £24,000 and £28,000 – too good to be true, surely?

The XK Leaps In

It is strange that the car turned out as well as it did in its looks and overall road behaviour, for it was never intended to make a sports car so soon. The new engine put into the XK 120 had actually been under development for a new Jaguar saloon, the Mk VII, but it seems that this was being held up because of problems in the body pressings. To introduce their brand new power unit in the not entirely modern Mk V would have been a bad piece of PR for the new engine. As it had already attracted so much publicity, general interest was now very high in Jaguar's latest engine and the public should not be disappointed. How would they get out of this dilemma?

Not many weeks before the 1948 London Motor Show, the idea was formed to put the new motor into a sports car, especially as it was, after all, a brand new thing and sports car owners were recognized to be more enthusiastic, and therefore more tolerant of any niggling teething problems. This would give Jaguar some very valuable feedback on the new engine in service. This was a fine idea, but there was hardly any time left.

A Mk V chassis was cut down, offering its brakes, transmission, suspension and steering directly to the project, while William Lyons set to work designing a mock-up body. The whole process of styling took an incredibly brief two weeks.

A first-class rebuild in progress. XK 120 chassis with engine and gearbox, awaiting its suspension. The strength of the chassis is plain to see from this view.

The XK Leaps In

Perhaps that may be one reason for the Lyons styling triumph, for he never lost sight of the contours he was after. Besides, fear of failure – in this case missing the deadline of the show – concentrates the mind wonderfully.

The intention as the show opened had been to make an initial run of a couple of hundred of the new two-seater with aluminium bodywork supported by an ash frame, steel still being somewhat scarce. This figure may seem very low, but the factory only delivered a total of some 330 versions of their previous two-seater, in production for a period of four years. However, before the 1948 show was even half over, it was quite obvious that they had underestimated the demand for the XK 120 and that 200 cars would be hugely insufficient – by about 12,000 in fact.

Impressed as the visitors were, the Earls Court show car had hardly been anywhere under its own steam, let alone been driven in anger. It has also been described as little more than a mock-up. This may possibly explain the advertised weight figure of 2,470lb (1,120kg) – about 440lb (200kg) short of reality. Indeed, recent research has discovered that it only had the 3.2 litre engine under the bonnet, the six cylinder example of the XJ unit. This was because the lack of low-speed torque was not decided to be serious enough to need a remedy until shortly before the show. The subsequent lengthening of the stroke gave the unit its 3,442cc capacity.

FURTHER DEVELOPMENT

In the months after its public appearance the show car, the original prototype XK 120, now registered HKV 455, underwent a continuous process of development. In fact, it was only now that the car's development really began. Many test drives were taken by R. M. V. Sutton – nicknamed 'Soapy', although we are not told why – and he would return after each one for a debriefing with Heynes. All manner of small problems appeared and had to be dealt with, such as wrong damping rates, tyres rubbing on the bodywork during hard cornering (this happened again on HKV 455 at Silverstone) and overheating in heavy traffic. Steering geometry came in for some experimentation and many other minor modifications were made.

There is nothing unusual in this state of affairs, as anyone who has substantially altered a car of their own will testify. No matter how sure you are that a modification will work perfectly, driving the car is the only way to reveal if you were right first time. Also, one change often leads to the need to change something else. This can go on and on, like falling dominoes. After much testing and development of this trial and error sort, the car was more or less ready to be put into production. Before that, some verification of its top speed had become a high priority.

One intriguing aspect of HKV 455 is its lights. When viewed at the 1948 Motor Show it was seen to wear smaller lights on its front and rear wings than those subsequently fitted to the car and to all other XK 120s – except for HKV 500. This sister car is seen in many early shots wearing those same tiny lamps front and rear that had been seen on HKV 455 and that had by this time graduated to the production items. The fact that the sidelights on HKV 500 are different can be seen in the line-up for the first Silverstone race where HKV 455 now has the production-sized items, as has the third XK in the race (670001, the first left-hand drive car). Since both HKVs were apparently painted bronze when made, could it be that HKV 455 and HKV 500

The XK Leaps In

There are always perks when you become a mayor. The first XK of all, chassis 660001, most probably sometime before the car's competition début at Silverstone in August 1949.

exchanged identities? It is more likely that the lights were swapped when it was decided that HKV 455 was going to be road tested by *The Motor*, and therefore production-style lamps should be seen on the car. By the time HKV 500 was painted for the third time, for the 1950 Silverstone International Trophy, it wore the standard chrome sidelights, but was still seen to be wearing smaller tail lamps a year after this.

While all this fettling was occurring, dreamers, would-be purchasers and those who had actually placed orders were becoming restless. Several months had elapsed since its appearance and still no XK was yet offered for sale. Much muttering went on in the press, not just by journalists, as a glance at the correspondence columns in the motoring journals will demonstrate. In fact, Jaguar would silence their critics soon enough with the run at Ostend. The car went rather faster than the claimed 120mph (195km/h) – a speed that many doubted could be attained anyway. In so doing they happened to break

The XK Leaps In

HKV 500 being used for promotional purposes at the Bugatti Owners' Club Meeting, Silverstone, June 1949. Note that this car has yet to be converted to right-hand drive and that it wears small front sidelights.

the record for the highest speed recorded for a 'standard production model'. This was indeed ironic, since the XK 120 was not yet in production.

The highest speed recorded on the Jabbeke carriageway is always quoted with absurd precision as 132.596mph (213.392km/h), the average of two electrically timed runs. Just about its only roadgoing rival for speed would be the pre-war Bugatti 57SC, which was no longer being made, was extremely rare and was a very much more expensive car than the Jaguar. However, it is unlikely that the 'showroom' XK 120 would prove to be as fast as this.

The speed was achieved with an undershield (sometimes called a 'belly pan') fitted – not a standard part, although it was available as an option to speed-conscious owners. Also, just as the first E-types in 1961 had been tweaked for a little extra power in order to reach that magic 150mph (240km/h) – D-type cam cover breathers can be seen on one road test car – those very first 120 roadsters were probably also 'breathed on'. Any manufacturer who supplied a sports car for a high-speed demonstration run and did not at least 'blueprint' the engine would be a very unambitious one – hardly a description that could ever be given to William Lyons.

It is true that the first XK 120 to undergo an independent road test managed a carefully recorded two-way average of

125mph (200km/h). However, this was HKV 455, which had been entered for, and led, the *Daily Express* Production Car Race at Silverstone in 1951. What, then, was the performance of a truly standard XK 120?

Early in 1950, the first steel production 120 was given the full road test treatment by *The Autocar*. Although this car, JWK 675, would go on to win the 1951 Production Car Race at Silverstone, its performance suggests it was a completely standard car at the time of the test, over a year before the race win. Although *The Autocar* did not actually try for the top speed, they estimated it to be around 115mph (185km/h). As for weight, 670185 was recorded as 2,920lb (1,324kg), which compares interestingly with the car road tested by *The Motor*, HKV 455, the first XK made, which was found to be 2,856lb (1,295kg). However, when *The Autocar* weighed 'one of the earlier aluminium-bodied cars' from the 1949 Silverstone race, they made their steel car lighter by 40lb (18kg). So it has to be assumed that all that ash framing needed to hold up the aluminium skin offset much of the weight saving of alloy over steel. Besides, the cars were not built from aluminium primarily to save weight and this fairly minor difference of about 65lb (30kg) would affect initial acceleration only very slightly and then perhaps not at all above 60mph (100km/h).

Weight has virtually no influence at all on top speed, which is all about three things: horsepower, size and shape. The journalist who eagerly announced that if equipped with the 3.27:1 (high ratio) back axle, the XK 120 would manage 140mph (225km/h) was not being very technically minded. Around 70bhp extra would have to found for this speed, assuming the car ran in its normal trim. Extra horsepower must surely be the reason for the increase of some 10mph (16km/h) in top speed of the alloy over the steel car, as well as a whole six seconds less for the 60–100mph time, where any weight saving, especially one of only 66lb (30kg), would be practically insignificant (see panel).

Formula for Maximum Velocity

Perhaps we can shed some light on the power needed to gain this 10mph (16km/h). Using the formula provided, the power output for a standard 120 is estimated to be around 115bhp DIN (Deutsches Institut für Normung, the rigorous German Standards rating.) It is quite normal for the DIN horsepower to be 75 per cent of the originally quoted SAE (Society of Automotive Engineers) gross figures. The later 265bhp 3.8 litre XK engine is now reckoned to have yielded somewhere between 180bhp and 190bhp DIN. It can also be calculated that the 10mph (16km/h) increase to 125mph (200km/h) would have required a further 27bhp, assuming the Cd remained the same.

Let
v = Max velocity (m/s)
Cd = Coefficient of drag
ρ = Max power at flywheel (w)
r = Approx density of air (kg/m^3)
A = Frontal area (m^2)

Use

$$v = \frac{2p}{\sqrt[3]{\rho C d A}}$$

Where
1mph = 0.447m/s
1m^2 = 10.765ft^2
ρ = 1.25kg/m^3
1bhp = 0.746kw = 746w

The result for v will need reducing by 8 per cent (so multiply by 0.92) to take rolling resistance and other factors into account. This correction remains fairly constant throughout the speed range.

The XK Leaps In

IMPRESSIVE PERFORMANCE

All the above arguments are pretty academic, for as far as speed went in early post-war Britain, even a completely standard new open two-seater Jaguar was easily King of the Road. However, there is rather more to a car than its maximum speed. How were the XK's other attributes received?

Almost as remarkable as the car's outright maximum speed performance was the flexibility of its engine: maximum torque was delivered at a very low 2,500rpm. This helped one magazine coax the car from a standstill in top gear to 100mph in a time quicker than any car tested previously – these using their gearboxes to the full. Truly amazing stuff, and this exceptional low-speed tractability would become

XK 120 Specification (1950 Steel Roadster)

Engine
Type	DOHC, 6 cylinders, 7 main bearings
Material	Chrome iron block, aluminium head
Dimensions	83mm x 106mm, 3,442cc
Induction	Twin 1.75in SU carburettors
Maximum power	160bhp SAE (120bhp DIN) at 5100rpm
Maximum torque	195lb/ft (gross) at 2,500rpm

Transmission
Clutch	10in single dry plate, mechanical operation
Gearbox	4-speed synchromesh gearbox (not first gear)
Final drive	Hypoid bevel, ratio 3.64:1 (3.27 optional)

Suspension
Front	Independent by torsion bars, double wishbones, telescopic shock absorbers and anti-roll bar
Rear	Live axle, half-elliptic springs and lever dampers

Steering   Recirculating ball, worm and nut

Brakes   12in drums, hydraulic operation, two leading shoe at front

Dimensions
Length	173in (4,395mm)
Width	61in (1,560mm)
Height	52.5in (1,335mm)
Wheelbase	102in (2,590mm)
Track – front	52in (1,295mm)
Track – rear	50in (1,270mm)
Turning circle	30.8ft (9.4m)
Frontal area	17sq ft (1.58m^2)
Coefficient of drag	0.47 (approx)
Weight	2,920lb (1,325kg)
Distribution	48/52

known as a feature of the XK-engined Jaguars generally. The main reason for this excellent torque characteristic is the combination of a two-valves per cylinder layout and the long stroke.

To put the XK 120's excellent low-speed torque into perspective, it may be interesting to note that the first 120 to run through a full road test was faster in top gear from 30–50mph than a 6 litre, V12 McLaren F1. Accepting that the McLaren is carrying a very tall sixth gear, even the later, more representative, steel 120s were faster in top over this speed range than, for example, a current Ferrari F355 and Aston Martin DB7. What this means is that you do not need to keep rowing the car along with the gear lever to make rapid progress and this makes for a car that is faster in 'real' road situations than through-the-gears performance figures suggest. The 0–60mph times of around seven seconds recorded by the early six-cylinder E-type are nothing special today, but its 50–70mph times in top gear are (McLaren missile excepted) still unbeaten by any of today's current production cars, TVR Griffith 500 included.

BRAKES AND GEARBOX

A fast car needs fast brakes, and by the standards prevailing in 1949 many road users would have considered the car's brakes to be fine. Road testers tended to

Mrs William Boddy trying the XK 120 for size. The front bumpers are slightly askew – hopefully nothing to do with the loss of stopping power.

The XK Leaps In

praise them, although William Boddy of *Motor Sport* must be congratulated for managing to reveal the brakes' shortcoming: their tendency to fade. Presumably he was the only journalist to drive fast enough to find out. The great problem was stopping from high speed, especially repeated stops. The author's 1950 steel car had excellent brakes – providing 60mph (100km/h) was not exceeded. This explains why the brakes turned out to be the great Achilles heel of the XK 120 when the cars were raced.

It was not so much that the car was under-braked – its linings were usefully wider than those of the SS 100, which was a stranger to brake fade – as an inability to prevent the heat build-up which had such a disastrous effect on retardation. Indeed, it is generally accepted that the XK 120 was the first production car to suffer from serious brake fade. The reason for this is simply its inability to dissipate the heat built up in its brakes, caused by a combination of the car's weight and performance.

Its 16in wheels, which were 2in smaller than the norm hitherto, left less room for cooling air to circulate. However, the six-cylinder BMW 328 – as desirable in its day as the XK 120 would become – also sported

The enclosed rear wheels did nothing for heat dissipation from the rear brake shoes. Note the bent rear bumper iron and poor fit of side panels.

The XK Leaps In

Not too much brake trouble to be found on a hill climb. Shelsley Walsh, August 1954.

16in wheels without braking problems. The answer to this lies in weight, for the BMW scaled an astounding 1,210lb (550kg) less than the Coventry machine. The solid steel wheels of the early XK 120s (as opposed to later wire spoke wheels), together with its all-enclosed bodywork, impeded heat dispersion from the overworked drums. The XK's high performance was also, of course, a major factor in its braking problems.

The rapid acceleration of the XK and its ability to regain high speeds so quickly after braking would mean that the brakes had insufficient time to cool. Had such acceleration been available to more prewar sports cars, chronic brake fade might have been experienced earlier. As it was,

The XK Leaps In

perhaps we should be grateful to the XK 120 and its fade-prone brakes, for its brake-induced failure at the 1950 Le Mans led its makers to become the first manufacturer to fit disc brakes for this race in 1952. Jaguar was also one of the first makers to fit discs to its road cars – these had first been used on a 1949 Chrysler, on all four wheels. The first racing car so fitted was also American: a 1938 Miller.

Apart from the brakes, the other great failing of the car was often said to be the gearbox. The driver's handbook for the XK 120 advises that the gear changes should be 'slow and deliberate', which seems rather an admission. However, early road tests for the 120 did not criticize the gearbox, although they would in time, of course. Perhaps the real problem lay in the fact that it tended to wear out its synchromesh cones rather effectively, so that by the time cheap XKs started to appear, their gearboxes had become rather slow in action, especially the first-to-second change. The author's experience with a carefully rebuilt Moss gearbox was that he could change up between any gears quite silently as fast as he could move his hand. A downward change needed the accompaniment of a blip on the throttle to get the two relevant gearwheels turning at the same speeds. With a lightened flywheel and straight-through

The much-maligned Moss gearbox. This is to be fitted to a left-hand drive car, as shown by the bearing bolted at the bottom of the clutch cover.

exhausts, this was no hardship at all. Long-term familiarity with the car may have much to do with this ease of slick cog-swapping: journalists have only a few days to attempt to find compatibility. Nevertheless, when the E-type appeared, its old Moss gearbox was very dated and really no longer good enough to be in the latest Jaguar sports car.

Even poorer in performance than the brakes, the headlamps of the XK 120 were really rather hopeless. They would not attract the amorous attentions of a love-starved glow-worm, even on main beam. That they were criticized quite harshly in the press on the car's introduction – when the brakes and gearbox generally were not – demonstrates just how poor these lights actually were. Although the headlamps were improved for the XK 140 and 150, which also had provision for extra spot-lamps, Jaguar unfortunately reintroduced their tradition of feeble lighting with the first of the E-types. At least it prevented you from going too fast at night.

RIDE AND HANDLING

The XK was equipped with Hassan's excellent independent front suspension. Its claim to excellence can be gauged from the

Not a lot of candle-power here: the XK 120 headlamp.

fact that it was still in production on the last of the E-types – some twenty-seven years after its introduction on the Mk V. The rear suspension was very conventional, almost old-fashioned, with its cart springs, lever dampers and complete lack of locating links. How did this set-up behave in terms of ride and roadholding?

The amazing performance of the 3½ litre car was not the XK's only sensation. The other was its comfort, achieved by a combination of ride quality, silence and luxurious interior trimming. With its generous, curved leather seats, its adjustable steering column, quality carpeting and complete absence of bare metal visible in the cockpit, it was really very plush when compared with the usual British sports cars of the period.

It would also be quite fair to say that the XK was the first ever comfortable-riding sports car. This was really the fourth sensation after the car's speed, beauty and price. The SS 100, for example, was far harsher, a real boneshaker by comparison, but no different in this respect from its contemporaries. The 'boulevard' ride, as it was often described, was allied to a standard of roadholding that enabled the XK 120 to win very many international races – at one time more than any other sports car, an accolade only surrendered in the 1970s to the Datsun 240Z. With its relatively crude rear suspension and its rear weight bias, you might expect the XK to be tail happy. *In extremis*, it will understeer, although a forceful right foot in low gear can turn this into oversteer – especially in the wet.

If the roadholding was good, the handling was less so. As one road tester once put it at the time, an XK can be cornered very fast, although it resents being asked to change direction suddenly. Perhaps a look at some of the photos of XKs racing will demonstrate a lack of finesse in the handling, from the amazing angles that the cars adopted in the corners. They are leaning noticeably more than other cars – which they still beat. The roll exhibited by these early racing XKs is almost to Citroën 2CV levels. Perhaps it would be fair to say that the 120 was one of those cars that felt much better on a corner from the driver's seat than it looked from the outside. It would not be long before the suspension was stiffened with thicker torsion bars and anti-roll bar at the front as standard, these having first been offered as options.

These criticisms aside, the fact that a lavishly appointed and softly sprung car like the XK 120 would prove faster than all its British sports car competitors in its first race demonstrates the lead that Jaguar had taken in the art of combining excellent ride and handling, previously held to be irreconcilable. Nearly fifty years later, a Jaguar is today still revered for its combination of a supple and silent ride, allied to an agility and grip that do not really belong in the large car class.

FITTINGS

The level of equipment was generous indeed. The XK 120 had a full array of Smith's instruments: oil pressure and water temperature combined in one gauge, an ammeter and a fuel gauge. At either side of these were a large 5in (125mm) speedometer and rev counter with white lettering on black that has yet to be bettered in appearance. Unfortunately, the dashboard that carried these instruments was set into the middle of the car and this could require quite a lengthy glance away from the road straight ahead. This central position for the instruments would prove a bonus, of course, when converting the car to or from left-hand drive.

The XK Leaps In

An XK 120 being cornered hard. To roll this much before sliding, it must be shod with radial tyres and be wearing its early type anti-roll bar.

A heater was not a standard fitment until more than two years after the car's announcement; slots for demisting the windscreen would take a further year to appear. Anyone who has driven an early XK may remember the frustration of trying to drive a car where one's breath froze on the inside of the windscreen. Drivers of air-cooled Volkswagens will find such antics familiar.

The steering of an XK 120 was far lighter than you would believe nowadays, with that large lump of chrome cast iron sitting above the front wheels. No power steering would be offered for the XK 120, of course, but the narrow cross-ply tyres of the time, coupled with that huge 18in (455mm) diameter steering wheel meant that the steering was easily manageable when parking and quite light on the move, although hard cornering would cause it to load up somewhat. Pumping up the tyres by around 10psi would make the steering of the 120 extraordinarily light and delicate, without too much harm done to the ride comfort, except on very poor surfaces. However, there was always a chance of wearing out the tyres in the middle by running them over-inflated. The turning circle was just incredible, being 20in (0.5m) tighter than Rover's 1995 100, its smallest car.

The imaginary XK 100 was turning out to be just that. It had a number of things going against it from the start. Conceived mainly for the thin wallets of post-war Britain, it seems that the departure of the 1½ litre saloon soon after the XK's introduction had made Jaguar think twice about building small-engined (four-cylinder) cars. Also, it had the usual vibration common to all four-cylinder cars and missing from the sixes (perhaps with the exception of the early Ford V6s). The fact that it was only a three main bearing design cannot have helped its smoothness. Furthermore, as its price was the same as the 3½ litre car, it was no wonder that nobody took much notice of it. (Did anyone actually try to buy one?) It just faded away, although it seems it was still on the back burner as late as 1953, possibly to power the new range of compact cars.

Although at least one XK 100 was made and was often seen, painted blue, driving around the works, it is good news that a 2 litre XK was never put into production. Jaguars with small engines have a history of being under-powered, without the benefits of good fuel consumption. The 2.4 was up to 25mph (40km/h) slower than the 3.8, took nearly 20 seconds longer to reach 100mph (160km/h) and was often almost as thirsty. Similar accusations can be levelled at the 2.8 and later 2.9 XJ6 models (the former of which used to seize up quite frequently into the bargain). The whole problem stems from asking a relatively small engine to power a relatively heavy car. The smaller power unit has to be worked much harder to extract any performance, so that the fuel consumption suffers. Besides, it would not have made much sense for Jaguar to offer a sports car that could not keep up with their saloons.

HARD TOP

A fixed-head coupé version of the XK 120 appeared in March 1951 and was, when viewed in profile, the finest looking Jaguar ever made. The way the curves of the rear of the roof, the rear side window, and the tail all come together is pure, sensuous artistry. The roadster has the beauty advantage from a full-frontal point of view, thanks to its prettier windscreen. The covered-in Jaguar weighed about 110lb (50kg) more than the original open model.

In September 1951, six months after its introduction, the fixed-head coupé became available in Special Equipment guise. Thus specified, it had stiffer front suspension thanks to thicker torsion bars and anti-roll bar, and 180bhp. A lightened flywheel would help low-speed acceleration; the 20bhp increase stemmed from a higher lift on the camshaft, and a straight-through exhaust system. It is unfair to accuse it of being a silencer, and it was officially known

The XK Leaps In

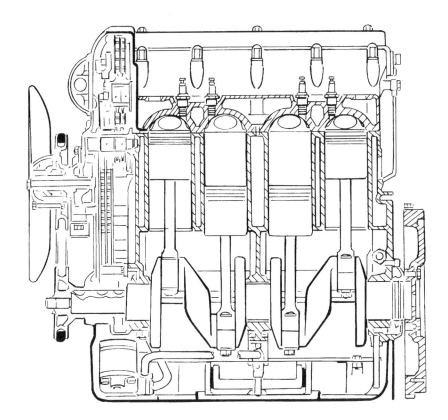

The 'mythical' XK four-cylinder engine that sat on the back burner for several years before being permanently shelved – thank goodness.

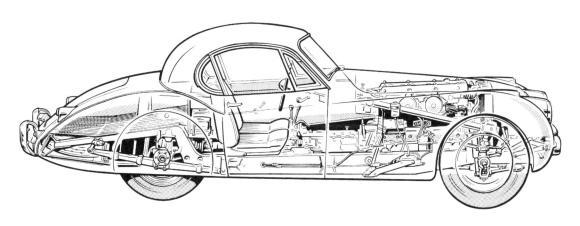

The XK 120 fixed-head coupé laid bare.

The XK Leaps In

as the 'twin pipe racing type', but was usually referred to as the 'competition' or 'C-type' exhaust. This gave a delightful deep-throated growl under load, becoming a bark at about 2,500rpm. Jaguar returned to a single pipe system on the fixed-head SE cars after about six months. Perhaps the noise was just a bit too much in the closed car.

Despite its power advantage over the standard car, as well as being slightly more aerodynamic than the open cars, when tested by *The Autocar* a Special Equipment fixed-head coupé just managed two miles a minute, although the magazine hinted that a little more may have been achieved but for the traffic present that day on the Jabbeke highway. However, this car still took longer than the suspiciously rapid HKV 455 to reach 100mph (160km/h), despite having a lower ratio in its back axle. This suggests that the top speed of a standard, off-the-shelf, 160bhp roadster XK 120 was really about 115mph (185km/h). (In other words, the 120mph-plus speeds of the first cars were reached not only by rudimentary attempts to lower the drag coefficient, but by tweaking the engine.)

Towards the end of 1951, the rapid expansion of the company meant that more space was once again needed. Once again, a factory with wartime connections was to be the new venue. This one was a result of the hurried rearmament programme in the

Leaning less than the first XK 120s, but the lines of this coupé driven by Jack Sears are compromised by the mascot sticking out from the graceful slope of the bonnet. Snetterton, October 1954.

The XK Leaps In

This eye-catching combination has earned no apparent interest from the blasé Parisian passers-by.

late 1930s and was termed the Number Two Daimler Shadow Factory. The move was done in stages – no longer could they move lock, stock and barrel over a weekend – and was completed the following year.

The fixed-head coupé XK was followed in April 1953 by a drop-head model. This weighed about the same as the tin roof car, but was not as attractive as either of its two close relatives, especially with the top down, when it tended to look a little like a pram. However, it was far easier to erect the hood on a drop-head than on the super sports, which needed skills normally attributed to Boy Scouts. Also, with its wind-up windows and hood with twin linings – one to hide the mechanism – the drop-head was a very much more civilized

The XK Leaps In

Fine car, but a miserable place to be in the rain. It is often worse with sidescreens erected.

place to be in the wind and rain than the cockpit of the roadster where, hood up, the sidescreens would flap wildly and the rain would always find a way in. A long and wet journey at speed in a roadster with the top up is no fun at all, unless you have a well-developed sense of masochism – just like the sort of person who enjoys camping.

With the range now extended to three body styles, albeit with the two convertible types overlapping rather in what they offered, there were still shortcomings in the XK 120. Overheating in traffic jams was not uncommon, those brakes and headlamps really did needed improving, body protection from American parking methods – where judgement of distance is replaced by the audible clash of fenders – was feeble, and leg room was cramped for anyone approaching 6ft (1.8m) tall. This last factor was a slight disadvantage in the American market, where people were generally larger.

These problems were addressed with varying degrees of success in the forthcoming XK 140. Although improved on this model, the braking system would only really become worthy of the car with the introduction of the XK 150. Extraordinarily enough, Jaguar would reintroduce headlamp, brake and leg room deficiencies with the 1961 E-type, which would also gradually come to lose its original beauty – just like the XK.

4 Stretching its Legs – The XK 140

The XK 140 was a better car than the XK 120, but less appealing. It was introduced at the Earls Court Motor Show in October 1954, six years after its forerunner's sensational first appearance. In chassis terms, it differed from its predecessor in only one obvious way: it had rack and pinion steering. So little was the XK series to change in engineering terms that the only other major change to affect the range would be disc brakes, introduced on the XK 150 in 1957.

Mr Denis Flather suitably attired for a night out with his XK 140 fixed-head coupé. If fitted with C-type engine option, it could pull around 125mph. In de-restricted Britain of the mid-1950s, this was just about unmatched.

Stretching its Legs – The XK 140

Visually, the XK 140 was at first glance the same the 120, except for its much heftier bumpers and somewhat clumsier grille. A longer look will reveal differences in details such as headlamps, front indicators, central chrome strip, rear lights, number plate mountings and boot lid handle. Inside the cockpit there were also many small changes, the most obvious being to the steering wheel boss and seats. Nearly the whole of the body was different, except for the rear wings. Despite the bonnet, boot lid and doors (early cars only) being fabricated in aluminium, weight unfortunately went up by an average of around 165lb (75kg) over the equivalent XK 120 models. The mechanicals had many small changes as well.

The essence of the bodily changes was that on the two convertible cars the engine and bulkhead were now moved some 3in (75mm) further forward. Changes to the

The XK 140 (and XK 150) headlamp. A certain improvement over the XK 120, but no match for a modern halogen item.

Stretching its Legs – The XK 140

cockpit also meant that more room was found rearwards as well as forwards, making it a far less cramped place for the taller driver. This was mainly achieved by resiting the batteries in the engine compartment, from their position behind the seats in the 120. This allowed some 3in (75mm) of extra seat adjustment and must have considerably aided sales in America. On the drop-head model, two very small seats were provided in the rear. If the main seats were not too far back, they could just about house two small, young persons, or an adult sitting sideways. (The author was once driven for around one hundred miles in this fashion in his own car, but had to be prised out at the finish.) The fascia as a whole was lifted an extra 1in (25mm), helping to give more thigh room under the steering wheel.

Similar changes occurred on the fixed-head, but this time extra changes to the bulkhead gave a further 2in (50mm) of interior space and this was used to provide slightly more sensible room in the rear seats, although it could still only be described as a 'two plus two'. The bonnet and tail on a fixed-head XK 140 were shorter than on the other two models. Compared to the 120 fixed-head coupé, this gain in interior room and practicality was offset by a loss of grace, particularly in profile: precisely the same thing that would happen to the E-type when the 2+2 appeared. The 140 fixed-head coupé is probably the least attractive XK – aside from an XK 150 with disc wheels and spats on the rear, fortunately a very rare sight.

SIZE AND SHAPE

The overall dimensions of the XK 140 were almost exactly the same as its predecessor,

XK 140 fixed-head coupé – the see-through version.

51

Stretching its Legs – The XK 140

The least attractive XK, although decent bumpers are clearly displayed.

except that the larger bumpers would account for a small increase in length and width. These bumpers were enough to mar the very fine lines of the XK 120, the point being that the apron joining them to the front of the car hides the bottom curve of the front wings, which remains visible on the XK 120. This means the XK 140 looks slightly squarer at the front, although the 150 with its dip in the bumper is not quite as bad in this respect. Furthermore, the one-piece casting of the XK 140 grille may have been a much cheaper proposition than the previous car's fourteen-piece construction, but it was far less attractive. The William Lyons styling masterpiece that was the XK 120 was gradually being compromised.

Although the forward shift of the engine meant that the car's weight distribution became almost 50/50, compared to the slight 48/52 rearward bias of the XK 120, moving the engine forward did not improve the handling and roadholding. After all, just about every racing car since the mid-1930s has had a rearward weight bias, this becoming more marked with the introduction of the mid- and rear-engined cars.

It is true, however, that a car with a frontal weight bias will be more stable in a

Stretching its Legs – The XK 140

cross-wind – hence the popularity of the front-wheel drive car – and certainly an XK 140 will run arrow straight at speed. (The author recalls demonstrating how he could take his hands off the wheel of his 140 at 100mph (160km/h), though the passenger was strangely unenthusiastic.) However, straight-line stability, roadholding and handling are not the same thing. Perhaps it would be true to say that for ordinary road use, the 140 was slightly easier to drive fast than the 120, but for ultimate roadholding in the right hands, the earlier car was the better bet. The large number of race wins that the 120 continued to amass *after the introduction of the XK 140* support this claim.

PERFORMANCE

Although the XK 140's engine was moved forward primarily to offer more space in the cockpit, this brought with it the added bonus of enough room to fit an overdrive unit, not possible on an XK 120 with the

Despite its slightly more civilized nature, owners of the heavier XK 140 still entered many motor-sport competitions with gusto.

Stretching its Legs – The XK 140

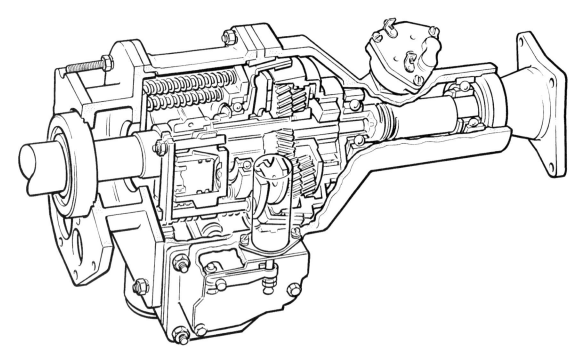

Overdrive unit, courtesy of Laycock de Normanville. Stronger internals can now be obtained for longer life.

Typical XK Axle and Gearbox Ratios

Axles

	Teeth on: C/wheel	Pinion	Ratio:1	Remarks
ENV	51	14	3.64	Standard early 120
ENV	49	15	3.27	High* ratio early 120
Salisbury	43	13	3.31	High* ratio 120
Salisbury	46	13	3.54	Last of 120s; 140/150
Salisbury	49	13	3.77	Later 120
Salisbury	45	11	4.09	140/150 with over-drive

*Strictly speaking, these are *low* ratios.

Gearbox

	1st	2nd	3rd
XK 120 and XK 140	3.375	1.98	1.37
XK 140(CR) and XK 150	2.98	1.75	1.21

great length of the rather aristocratic-sounding Laycock de Normanville overdrive then available. This meant that an XK 140 could be bought with five forward speeds, which in turn meant that while the final drive could be lowered by over 15 per cent over the 120 to provide better acceleration (if using the same four gearbox ratios), a usefully higher overdrive top gear would result – over 12 per cent higher than on the XK 120.

The overdrive ratio was exactly 7/9 and would operate only on top gear, thanks to the wiring arrangement to the solenoid that went via a switch mounted on top of the gearbox. (This switch was identical to the one that operated the reversing lamp when reverse gear was engaged.) Control of the power to the switch on the gearbox – and thence to overdrive if the car was in top – was by a small driver-operated switch built in to the side of the fascia.

Unlike some other cars, overdrive on a Jaguar would only work on top gear, because of the damage that the torque multiplication of a lower gear might inflict on the epicyclic gears that formed the actual 'over-drive'. For example, 213lb/ft from the 'C-type' engine fed through an engaged overdrive while the car was in bottom gear, ratio 2.98:1, becomes 635lb/ft. Even third gear at 1.21:1 produces another 58lb/ft of twisting force with which the overdrive's internals have to cope. All the engine's torque in all gears (whether multiplied or not) goes through the one-way roller clutch at the end of the overdrive unit. When these are part worn they are not over fond of full power, particularly from the later triple-carburettor 3.8 unit. (The author once shattered a roller clutch while trying to accelerate up a steep hill in first gear.) It is surprising that overdrive was offered with the heavy 3.8 Mk X, because of the great strain on its roller clutch.

Overdrive on the XK 140 brought two great benefits. First, fuel consumption on a long run was markedly improved. The fixed-head coupé tested in 1955 by *The Autocar* managed 22mpg (12.9 litres per 100km) overall, the lowest fuel consumption recorded for any XK model. Furthermore, high speed cruising was more relaxed – being achieved at lower engine revolutions, which also brought a reduction in engine wear. The long-legged gait afforded by the fitting of an overdrive to the XK 140 is one of the great pleasures of XK motoring.

The extra interior room of the XK 140 over its predecessor would help broaden the sales appeal of the car, now that it was starting to be less of a sensation in performance compared to its rivals. The XK was, after all, now entering its seventh year, and all manufacturers seek to enhance the appeal of a car as it gets older without the enormous expense of a costly redesign of the mechanicals. Bodywork face-lifts are cheaper than new engines and suspension layouts: a fact as true then as it is now. As for the chassis, this remained largely unaltered, except for the steering and the rear shock absorbers, which were now telescopic.

It is often pointed out that the XK chassis was immensely strong. It may have been when judged by the standards of its time. Certainly those side members are exceptionally deep – they would not look out of place on a box-girder bridge. The chassis was, after all, designed for the Mk V, which was 770lb (350kg) heavier than the XK. Being shortened would also make it more rigid still and since the XK's structural strength came in its chassis, it made little difference whether in open, drop-head or fixed-head mode. However, judged by a proper monocoque construction (with a roof) it would come a poor second. If this is

Stretching its Legs – The XK 140

The rear part of the XK chassis, being made deliberately flexible where it carries the rear springs, is therefore of smaller section. It is also more prone to rot at this point than elsewhere.

doubted, try lifting an XK under a rear bumper and you will see the rear of the door and its shut-face move relative to each other. Part of the reason for this flexing at the rear is the slender rear chassis members that carry the rear springs.

A roofless monocoque, like the E-type drop-head, is no more rigid than an XK. Modern saloon cars that are made into convertibles can be almost disgracefully flexible. The author recalls driving one very popular version from a major manufacturer which flexed so much that not only was the roadholding greatly affected but on a bad bump the scuttle shake caused the instruments to become blurred. Monocoque may mean 'single shell', but does an egg shell retain any strength after its top has been sliced off?

The steering of the XK 140 changed to rack and pinion, courtesy of Messrs Alford and Alder, and was mounted on steel and rubber sandwiches that were, in turn, bolted to the chassis. This new mechanism gave a slightly inferior lock to the 120, but a very slightly quicker ratio. It also allowed slightly more kick-back than the Burman recirculating ball steering on the earlier

Stretching its Legs – The XK 140

car, although a reduction in the castor angle on later cars would alleviate this somewhat. Rhoddy Harvey-Bailey, who must know as much as anyone about driving XKs, as well as setting them up for competition use, reckons that the steering of the XK 120 generally had a better feel than the rack and pinion of the XK 140. It is no coincidence that Mercedes-Benz waited until 1995 to offer their first road car with rack and pinion steering.

Recirculating ball and nut steering mechanisms can have a very pleasing feel, and rack and pinion certainly offers no improvement in this respect. Its attraction for any manufacturer is its reduction in moving parts, which makes it cheaper to produce. Having used a rack on the C-type and D-type, where weight saving would have been the aim, Lyons would hardly have hesitated at the prospect of saving money on a bought-in component. However, where the driver does gain from a rack and pinion system over the previous recirculating ball, is in the slight reduction in lost motion between the driver and the wheel, there being fewer joints to accommodate any play as well as less inertia in the fewer moving parts between the driver's hands and the steering arms bolted to the front uprights.

The rear telescopics did not seem to make that much difference when new, but experience would reveal that they kept up

The two wishbones of the front suspension: upper on left, lower on right. The torsion bar is splined into the collar at top right, the shock absorber is mounted on the shiny pin at bottom right, and the anti-roll bar is attached to the eye between the two.

Stretching its Legs – The XK 140

their performance rather longer than the lever ones. The XK 140's front suspension would now be the same as the Special Equipment option for the 120. The anti-roll and torsion bars would be of slightly larger diameter, increasing stiffness by about 20 per cent. The brakes were improved, thanks to more fade-resistant friction material and, more importantly, they were now self-adjusting at the front. Because of the increase in weight, however, overheating brakes could still be a problem if the driver began to get frisky, although the onset of fade was not quite so easy to induce as on the XK 120.

The engine in the XK 140 now incorporated as standard many of the Special Equipment modifications of the XK 120, such as high-lift camshafts, although it now yielded 190bhp (gross) compared to the 180bhp of the previous car, mainly thanks to superior inlet porting and adjustment of the ignition advance curve. This latter tweak would today be achieved by 'chipping'. The Special Equipment version of the XK 140 boasted a power output of 210bhp gross. This was reached by adding to the improvements already carried out on the power unit of the standard car. Enlarged ports, larger inlet valves and a twin-pipe exhaust system helped to find the extra 20bhp.

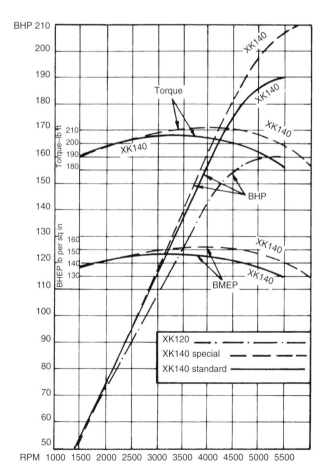

The C-type engine modifications available to the XK 140 gave a very useful increase in maximum power, but you had to use plenty of revs to get the benefit.

Stretching its Legs – The XK 140

The gearbox was still the rapidly ageing Moss unit from the 120 and apart from the options of overdrive and close ratios, it was the same as in the earlier car. Its change quality was now considered rather average – not surprising since the design was now more than ten years old. Would Jaguar, let alone its customers, have believed that this gearbox would be called upon to power two more future sports cars and not be replaced for another ten years?

An XK 140 specified with the Special Equipment options now had the power to take proper advantage of the extra gear and could reach higher speeds than the XK 120, if not actually live up to its type number as had that car in its later forms. One magazine recorded almost 130mph (210km/h) for an XK 140 fixed-head coupé, giving the speed attained to two decimal places. The absurdity of this precision is highlighted by the fact that the very same car was nearly 8mph (12km/h) slower when tested by another publication.

Nevertheless, with its excellent high-speed stability and overdrive gearbox, the XK 140SE was a better, and more economical, long-distance traveller than its

The XK 140's bumpers were not attached directly to the front wings, as on the XK 120. However, this shot shows that the brake cooling holes have effectively been reduced. That the 140's brakes were no more fade-prone that its forebear's may mean these holes achieved little in the way of brake cooling anyway.

Stretching its Legs – The XK 140

predecessor. It was roomier, more practical, more sensible, although rather less exciting. Perhaps it had grown up? If the XK 140 was a more responsible version of the youthful and impetuous XK 120, then the XK 150 was the same car fully in the flower of contented middle age. With the addition of chromium plating and those far more effective bumpers, its appearance had suffered, but the Americans loved it, buying XK 140s at a faster rate than either of the other two models.

XK 140 (1955 Fixed-Head Coupé, Special Equipment with C-Type Head)

Engine
Type	DOHC 6 cylinders, 7 main bearings
Material	Chrome iron block, aluminium head
Dimensions	83mm x 106mm, 3,442cc
Induction	Twin 1.75in SU carburettors
Maximum power	210bhp SAE (155bhp DIN) at 5,750rpm
Maximum torque	213lb/ft (gross) at 4,000rpm

Transmission
Clutch	10in single dry plate, mechanical operation
Gearbox	4-speed close ratio, synchromesh gearbox (not first gear), with overdrive
Final drive	Hypoid bevel, ratio 4.09:1

Suspension
Front	Independent by torsion bars, double wishbones, telescopic shock absorbers and anti-roll bar
Rear	Live axle, half-elliptic springs and telescopic shock absorbers

Steering   Rack and pinion

Brakes   12in drums; two leading shoe and self-adjusting at front

Dimensions
Length	176in (4,470mm)
Width	64.5in (1,640mm)
Height	55in (1,395mm)
Wheelbase	102in (2,590mm)
Track – front	51.5in (1,310mm)
Track – rear	50.5in (1,285mm)
Turning circle	33ft (10.1m)
Frontal area	17.1sq ft (1.59m^3)
Coefficient of drag	0.47 (approx)
Weight	3,130lb (1,420kg)
Distribution	52/48

5 More Comfort, More Power – the XK 150

The XK 150 was an XK 120 that had been sophisticated – in the true meaning given in the *Oxford English Dictionary*. Fixed-head and drop-head coupé versions of the new XK were announced in May 1957 and bore a strong family resemblance to the compact saloon, retrospectively termed the 'Mark One'. This new XK weighed as much as 330lb (150kg) more than the first XK; indeed, the fixed-head coupé 150 was found to weigh slightly more than the new 3.4 saloon, announced a couple of months pre-

It is a handsome car from this angle, where its new grille and dip in the bumper regain some of the XK 140's lost panache.

More Comfort, More Power – the XK 150

viously. That the two-seater was heavier than the four- or five-seater car is both a good advertisement for unitary construction and a poor one for body-on-chassis manufacture.

Not only was the XK 150 now the heaviest XK of all, but it looked it. This was mainly caused by the higher scuttle and waist-line that gave a very welcome extra width of some 4in (10cm) to the cockpit. The front screen was now a very attractive one-piece curve on both the coupés offered (no roadster just yet) and the rear window on the fixed-head coupé was much larger than on the previous XK model of that type. The extra leg room found on the XK 140 fixed-head coupé by extending the scuttle either side of the rear of the engine, was now found on both these new cars. The dash panel was now covered in leather on both models, and while the central instrument panel on the fixed-head coupé was also leather-covered, where the previous fixed-heads had walnut, the new drop-head was announced with an anodized aluminium instrument panel, as on the early E-types. However, very few of these seem to have been produced, the production drop-heads having the same covering as the fixed-head coupés.

This side profile is greatly enhanced by the whitewall tyres – although few cars in Britain were thus specified by their owners.

More Comfort, More Power – the XK 150

An unusual view showing the usefully larger rear window of the XK 150 fixed-head coupé. This is a late model with the larger rear lamps that necessitated moving the overriders in a touch.

In comparison to the XK 120 and 140, the 150's instrument panel looked as if it had some instruments missing. The leather covering and padded roll along the top of the fascia (and the windscreen rail in the drop-head) were the first attempts at passenger crash protection in an XK model.

The XK 150 was now a much easier car to fit in and to see out of, but the extra weight had its effect on the handling and to a lesser extent on the roadholding. Try putting three bags of cement in your car and then driving as fast as you might normally on some tight corners. This is not really recommended, of course, and in motor racing terms, even a disadvantage of 220lb (100kg) is a killer, as the BMW teams will tell, you as regards the British Touring Car Championship. They were forced to carry ballast of this amount to compensate for the advantage of being rear-wheel drive cars in a sea of front-wheel drive ones.

More Comfort, More Power – the XK 150

The dash now had a padded roll along its top edge, as had the two roadster models of the XK 120 and 140. The way the door panel has been reshaped to increase cockpit width can be clearly seen.

The most surprising thing about the XK 150 is the fact that it is the same width as the previous car, for it always appears to be much wider. Although its shape was definitely considered a more modern one than the XK 120 and 140 at the time of its launch in May 1957, it certainly lost its panache faster. It was the only Jaguar sports car offered with the leaping mascot, from the saloons, as an option. (A mascot mounted on that beautiful, sloping XK 150 bonnet added nothing to the car's already compromised grace. Fitted retrospectively to an XK 120, it did as much for that car's sleek shape as a wart on a nose.)

This restyling was pure common sense by Jaguar. It was also cleverly arranged so that it could be done without major retooling – trust Sir William (he was knighted in 1956) to avoid spending money if he could avoid it. 'Common sense', because when a sports car is getting a little long in the

More Comfort, More Power – the XK 150

The much maligned rear seats of the fixed-head coupé could occasionally be very useful. Sitting sideways, a none-too-large adult could travel reasonably comfortably.

tooth, a good way to boost its sales (or at least prevent a decline in them) is to make it more useful, more practical. This is why the XK 140 became more civilized and the XK 150 would also go along this route, only further.

A sports car as outstanding as the XK 120 most certainly worked wonders for Jaguar's prestige all around the world and this, in turn, brought in more sales than ever – of all the Jaguar range. In sales terms only, that is, ignoring the advertising that it did for Jaguar, the XK 120 was hardly the main breadwinner for the Coventry company. About 12,000 of this first XK were sold, as opposed to around 30,000 saloon models over the same period. This means the XK 120 made up about 28 per cent of the total production. By the time of the XK 150's introduction, it was no longer able to bring home the competition victories as the 120 had done – the D-type was now the car flying the flag at home and abroad for Jaguar – and so the XK 150 had to work harder for its living. This explains why it was designed to have a much broader appeal. Had it remained as starkly impractical as the first open two-seater XK 120 had been, sales of the XK 150 would have been far lower than they were. Despite the XK 150's better brakes, the basis of the car (engine excepted) was hardly very advanced from when it first appeared in the XK 120 of 1948, and now the design was almost ten years old.

Thus, to interest people enough to buy the XK 150 it was necessary to take some of the sport out of the sports car and lean towards the family car. The E-type 2+2 illustrates this strategy well. Perhaps the XK 150's looks belied its performance a little, for it was, even as it first appeared, a faster car than either of its forebears. Beauty is in the eye of the beholder, of course, but nowadays the XK 150 fixed-head coupé appears bulbous, even frumpish, if viewed beside the same XK 120 model.

Initially the XK 150 was offered with either the normal 190bhp XK 140 engine or a 'blue top' 210bhp unit, with its B-type cylinder head. The basic model with the lower power engine was available with disc wheels, although, unsurprisingly, it seems that just about every one of the new sports cars was ordered with wire wheels and the 210bhp power unit. Apparently, the basic car could also be bought with drum brakes, but it is not clear why anyone should want them, let alone why they were offered.

More Comfort, More Power – the XK 150

Oh dear. The possibly unique sight of an XK 150 roadster with disc wheels.

More Comfort, More Power – the XK 150

That the top speed of the 210bhp Special Equipment XK 150 was inferior to the XK 140 with its C-type 210bhp, was due mostly to the slight increase of around 6 per cent in frontal area of the newer car. In fact, it was only very slightly faster than the new 3.4 saloon. However, its acceleration was altogether stronger than even the XK 140 with the C-type cylinder head. This head may have produced the same maximum horsepower and very nearly the same maximum torque as the new B-type head on the 150, but since the peak torque of the new car was delivered 1,000rpm lower down the scale than the older design, the XK 150 was noticeably superior in acceleration, both outright and in the mid-range. This was because the B-type head was, basically, endowed with the larger valves of the C-type head, but kept the narrower inlet ports of the standard XK 140 unit. This would improve the low-speed torque to make the 150 usefully quicker in all normal driving conditions.

The long stroke of the XK power unit provides good leverage around the crank at low engine speeds. Also, the two-valve layout tends to keep up the gas velocity into the cylinder at these low engine speeds. The now fashionable four-valve layout is better for breathing at high crankshaft speeds and will ultimately give more power, but it is inferior to the two-valve layout at low speeds. This is because the cross-sectional area of the four-valve inlet porting is effectively larger where the gas enters the cylinder, thereby reducing the speed of the incoming charge, in comparison to the smaller throat of a two-valve head. It is this restriction that causes the mixture to speed up through the valves. Perhaps a good analogy is that by reducing of the size of a fire grate by blocking it off with a sheet of newspaper, you can make the fire draw much more strongly up the chimney. Useful as this improved torque characteristic might have been for acceleration, Jaguar's two-seater had one much greater advantage over its predecessors: it now had high-speed brakes.

Necessity is the mother of invention, as the proverb tells us, and it was Bob Dylan who told us that there was no success like failure. It was the general failure of Jaguar brakes under racing and fast road conditions that led the company, in conjunction with Joe Wright of Dunlop, to develop the disc brake before others in the motor industry. (Frederick Lanchester, possibly the cleverest person ever connected with the motor industry, had put a form of disc brake on one of his cars in 1902.) Jaguar and Dunlop had been collaborating on disc brake development since 1951 and as early as 1958 they had a Mk VII with anti-lock brakes – alas, never put into production. As a result of these efforts, the brakes on the XK 150 were, if the performance was going to be used at all, a great improvement over the previous model, and a huge advance on those of the first XK 120s.

The method of operation of the disc brake is common knowledge today. The reason it beats the drum, so to speak, lies in its ability to dissipate heat far more effectively than the all-enclosed drum brake. This is possible because the disc brake – as fitted to the first XK 150 – had a friction lining/swept area of 6 per cent, whereas this proportion on the drum brakes of the XK 140 and 120 was 61 per cent. This means that there is a much greater proportion of the brake disc that has time to cool while not in contact with the pads than has the drum when not in contact with the linings. Furthermore, the heat-generation source is enclosed inside the drum and therefore the heat has to find its way out – unlike on the disc, which is exposed.

More Comfort, More Power – the XK 150

A newly fitted XK 150 front brake disc. You can see that only a small proportion of the brake's surface is in contact with the friction material, in contrast to the drum brake.

One difficulty encountered when designing the disc brake was that the far greater combined volume of its wheel cylinders, compared to those for drum brakes, meant that much more fluid would have to be displaced if the pads were to move as far as a brake lining. This would have necessitated a pedal stroke of a half a metre or so, and so a very short travel was designed into the pistons of the disc brake. This was one of the main engineering problems and was overcome by the use of the extractor pins that keep the pads at a distance of 1/100in (0.25mm) from the disc.

One slight drawback of the early XK 150 disc brake concerned the changing of pads. These cars were equipped with round pads and this meant the calliper had to be removed to extract the worn pads. Later cars would have square pads, introduced with the S models, which could be slid out from the calliper after removing a small bolt. Perhaps the early cars' pad change was not so bad, since it would only have to

More Comfort, More Power – the XK 150

be done every year or two and would at least give a chance to clean out the calliper and carefully inspect the pistons.

Another problem was that the disc brake, being flat over its operating surface, has no 'wrap around' effect common to the curved operating surface of the brake drum, which can be turned into a self-servo action when a two-leading shoe arrangement is employed. Thus for the disc brake system on the new XK some assistance was given to the driver by means of a vacuum servo, in this case by Lockheed. With the servo completely inoperative, you can still stop an XK 150, but it is a very effective way to experience real fear. A small tank, which was rather prone to rust through on all Jaguars, was provided on later cars as an emergency vacuum store if the engine stopped. Almost all modern cars have brake servos and none, to the author's knowledge, carries a vacuum tank. The servo itself has enough reserve of vacuum for at least a couple of assisted stops.

When these brakes were being developed, the car was subjected to thirty 0.7g stops from 100mph (160km/h) with a 60-second interval between each. The pedal pressures were the same at the end of the test at the beginning. An XK 120 could probably not have accomplished two such retardations without total fade: this was very impressive progress by Jaguar.

This is not to say that disc brakes do not fade, for a motor race will soon reveal shortcomings in a braking system that no amount of reckless road driving can. The author was once involved with a factory-standard XJS that entered a club race at Snetterton. Here was a car whose brakes were universally praised in road tests, but after just two laps in practice its brakes ceased to work at all.

Certainly no road tester ever found any fade in the XK 150 brakes, but for emergency stops at urban speeds when cold they were sometimes not quite as effective as the old drum brakes on the 120 and 140. None of these cars could match the stopping power from 30mph (50km/h) of the venerable SS 100. Alas, the first E-types would have the poorest low-speed braking of all the Jaguar sport cars, but that is another story. Apart from urban-speed crash stops, hopefully not a common occurrence, there was one aspect in which the new disc brakes were unquestionably inferior to drums. This was in the strength of the handbrake, which would be the bug-bear not only of Jaguar handbrakes, but of disc brake handbrakes in general, for many years after.

Whereas a drum handbrake uses the same area of rear linings as the footbrake, the Jaguar disc brake handbrake (a fly-off type in the XK 150) utilized two small pads that gripped the disc in its own cable-operated calliper. The drawback of this system is that the area of friction material employed for the disc brake handbrake is about 10 per cent of that in a drum brake handbrake mechanism. It is true that the XK 150 footbrakes are some 15 per cent of the friction lining area of the XK 140 footbrakes, but the discs have servo-assistance to multiply the force applied by the driver, while the handbrake itself does not. In service, if the handbrake on a disc brake Jaguar is reassembled with the correct amount of pre-load, and the handbrake callipers are checked for signs of seizure, a great difference will be noted. The author has seen an XK 120 with XK 150 discs achieve 0.42g retardation in the MOT test, its handbrake having been set up exactly as the makers intended.

One of the problems that the XK 150 had to endure from its introduction was the fact that there was another car in the Jaguar family that not only was merely two or

More Comfort, More Power – the XK 150

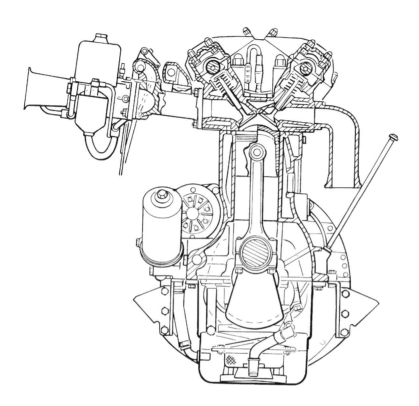

The XK 150S engine, with one of its three larger 2in carburettors visible.

three miles per hour slower in top speed, but that would take less than one second longer to reach 100mph (160km/h). The presence of the four-door, five-seater 3.4 saloon must have been a bit of a blow to the XK 150's sports car machismo, but in March 1958, the optional 'gold top' S-type cylinder head was offered, which would give the car a healthy boost in gross brake horsepower, up to 250, thanks to its three 2in carburettors, higher compression ratio and straight porting, courtesy of Harry Weslake. The original XK 150 would, of course, continue in production alongside the more powerful XK 150S version.

For some reason, the XK 150S is usually treated as a different car from the XK 150 when model totals are calculated. This is misleading, since the XK 150S is merely an XK 150 with a more powerful engine, and the Special Equipment XK 120s were not totalled separately from the standard 120, nor the 140 with the C-type cylinder head option from the basic XK 140.

At nearly 73bhp (gross) per litre, the 3.4S engine could boast the highest claimed specific output for an XK production engine, for as the engine was enlarged to 3.8 and then 4.2 litres, the laws of diminishing returns would apply. This was impressive progress from the 46bhp per litre of the XK 120 engine, in production only four years previously, for this represents an increase of almost 60 per cent. It is interesting to note that only two years later, in 1960, an XK engine would be able to claim 98bhp per

More Comfort, More Power – the XK 150

An XK 150S is, from most angles, indistinguishable from the standard-engined car. To modern eyes, those tyres look almost dangerously skinny.

litre, albeit as developed for racing. This would be the power unit in E2A, the Le Mans parent of the E-type. Its 295bhp from 2,997cc (at 6,800rpm) was a useful advance in specific output for a Jaguar racing engine, since the 1957 D-type needed 3,781cc to find almost the same power. Unfortunately, the reliability of this alloy-block 3 litre motor would never begin to compare with that of the 3.4 and 3.8 litre XK engines.

At the same time as the introduction of the S engine option, a roadster was added to the two coupés. The new open two-seater broke with Jaguar tradition and was offered with wind-up windows. Many

More Comfort, More Power – the XK 150

considered the car had 'gone soft' because of its new civilized image, but perhaps those critics had never had the pleasure of a trip in an XK 120 or 140 roadster with the hood up. Anyone over 6ft (1.8m) would have to duck down to look right or left and the way the side-screens flapped about at high speed was as annoying as it was deafening. With the arrival of the XK 150 open two-seater, the distinction between the roadster and the drop-head was shrinking. As time moves on, motoring history has shown that people want more and more comfort; the first XKs had no heaters, after all. The trouble was, the Jaguar roadster had now civilized itself out of a job and it was discontinued on the next range of sports cars. The E-type was offered only in

The XK 150 roadster. This particular car looks pleasingly neat, without the usual additions of wing mirrors, spot lamps and the wretched mascot.

More Comfort, More Power – the XK 150

No sidescreens to flap about on the XK 150 roadster. The large boot-lid strut would soon be replaced by a spring arrangement.

two versions – a fixed-head coupé and a drop-head coupé. This latter was called a roadster, however, presumably in an effort to make it sound a little more hairy-chested.

The XK 150S range in 3.4 litre guise had class-leading performance – excluding Ferraris, which were up to three times the price anyway. Its acceleration compared very closely to that of the C-type, a racing car built in very small numbers for Le Mans, and road tested just five and a half years before the 150S appeared. Racing certainly improved the breed's performance, for this more powerful XK version could comfortably exceed 130mph

73

More Comfort, More Power – the XK 150

This car, a 3.4 litre S model, showed up the variations in journalists' driving techniques by recording very different acceleration times for each publication.

(210km/h). Three weekly journals tested the same car, XDU 984, and managed three rather different 0–60mph times, the worst taking no less than 20 per cent longer than the best. This illustrates how different drivers will produce different results: you should not take individual figures published for acceleration times as written in tablets of stone. (It is presumed the car was not going 'off-tune', as the top speeds varied by only 1.5 per cent from the slowest to the fastest.)

By today's standards, the dash to a 60mph time of around eight or nine seconds for the 3.4 XK 150S is merely hot hatchback stuff. However, it could haul

More Comfort, More Power – the XK 150

itself from 60 to 100mph in direct top (all the S models would have overdrive) some eight seconds faster than a Renault Clio 1.8 RSi being thrashed through its gears. It was found that a 3.4 150S could accelerate from a standstill to 100mph (160km/h) in about 33 seconds using direct top gear only. This is the same time that a 1993 Ford XR2i takes to reach the same speed using, once again, all its gears to the full. Such comparisons may be a little contrived, but you can see that once the Jaguar has pulled its great bulk up to around 50mph (80km/h), these smaller cars are effortlessly reduced to specks in its rear-view mirror. One can see the considerable attraction, even today, of owning such a powerful, long-legged sports car.

Once again, however, Jaguar would produce a four-door saloon with performance to challenge its own sports car's virility. The 3.8 Mk II, introduced with the new Mk II range in October 1959, would prove faster than the standard 210bhp XK 150 and would run a standing quarter of a mile (400m) in virtually the same elapsed time as the 3.4 XK 150S. This time, however, XK 150 honour could be upheld by the announcement, at the same time as the new Mk II range, of the 3.8-engined version of the XK 150S. Only tested by one magazine, its performance would put it well out of reach of the fastest of the Mk II saloons, reaching 120mph (195km/h) before the four-door car could manage 110mph (175km/h). This 3.8S unit delivered a gross figure of 265bhp and was the engine that would eventually power the first of the truly legendary E-types.

Apart from the braking system, the chassis and suspension of the XK 150 was virtually the same as the XK 140, and therefore the XK 120. To prevent steering kickback offending the driver, the steering column now carried a rubber universal joint. While it is true that probably 19 out of 20 owners could not have felt any difference at all, it does show that the XK 150 was

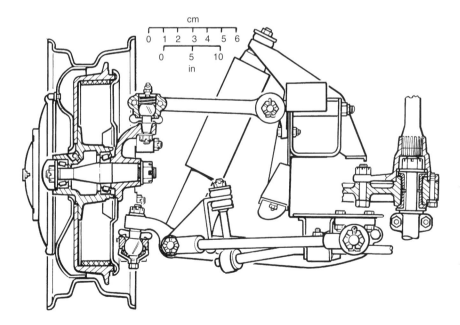

The only visual difference in front suspension between the XK 120 version (shown here with its drum brake) and that of the XK 150, is the top shock absorber mounting.

More Comfort, More Power – the XK 150

designed to take some small steps towards comfort and to tip-toe away from the ultimate in precision. (One of the first things to do when racing a road car – rules permitting – is to take the rubber out of the suspension and steering.)

A Thornton Powr-Lok differential was fitted to the S models, this preventing one wheel spinning fruitlessly on a full-throttle take-off out of a tight corner in first or second gear. There are mixed feelings about a limited slip differential when fitted to a poorly located live axle, such as on an XK. In the wet, the combination of XK 150S power in bottom gear, together with the relative crudeness of the rear suspension, meant that both wheels would spin wildly and the car would behave like a windscreen wiper as it slid sideways across the road. This could be quite a vicious thing, sometimes calling for armfuls of rapidly applied opposite lock. The other reason for misgivings as to the limited slip differential is that, once again, the dated XK rear suspension meant that the limited slip differential was overworked and so wore out its friction plates relatively quickly. Another suspension problem brought into the open by the extra urge supplied by the gold-top engine was spring wind-up during a fast start.

When the torque is unleashed and the wheels grip the road, the axle case reacts by rotating in the opposite direction along its axis and the nose of the differential is lifted, causing the rear springs to try to form an S-shape. Wheelspin starts for a moment, lowering the reactive rotational forces on the axle. The nose of the axle now drops allowing the spring to unwind. The wheels grip again and the same sequence is repeated. The frequency of this is something like 300 cycles per minute. With the extra grip afforded by modern tyres, as well as the fact that the Powr-Lok is making both wheels grip, this axle tramp is quite violent and causes a very loud juddering, even banging, until the accelerator is released. This rather ungainly mode of leaving somewhere in a hurry can be embarrassing, even expensive. (The author once caused a bout of axle tramp so fierce that the engine jumped forwards far enough to put the fan through a new radiator.) The cure for this problem, apart from driving somewhat more intelligently, is to fit tramp bars.

This was all proof that the rear suspension, with its complete absence of axle location (apart from the rear springs) was now rather out of date. This lack of constraint, which allowed the XK's rear axle to move around laterally, also had its effect on the roadholding, which was now generally conceded to have been overtaken by more sophisticated designs. However, the 150 was far from disgraced in this department, even if, as in all XKs, a bumpy corner met at speed would cause the driver's senses to go on red alert.

However, the car offered a fine ride, for the XK 120 really was the first comfortable-riding sports car and a bump will always deflect a heavier car less than a lighter one, providing the suspension travel is not all used up first. Helped by its greater weight, the XK 150 disturbs its occupants even less on the rough stuff than its two XK ancestors. The XK 150 roadster is also a far more comfortable car to travel in at speed, from the wind-protection point of view, than either the XK 120 or 140 open two-seaters. This said, with the XK 150 roadster being a particularly fine-looking car, perhaps better-looking than even the XK 120 version when viewed from the three-quarter rear view, it is, in its most powerful S form, a serious piece of kit indeed.

Another very significant plus for the XK's suspension generally was the forgiving

More Comfort, More Power – the XK 150

An early roadster with its overriders directly in line with the smaller rear lamps.

nature of the car when faced with clumsy driving. If the tail does come round, a snap lift-off of the throttle and some opposite lock will simultaneously retrieve the situation and flatter the driver's skill. (In Jaguar's next two-seater, with the large quantity of rubber in its independent rear end, such treatment would usually cause you to leave the road.)

This all-new Jaguar sports car would have a completely different rear suspension as well as a completely different body and chassis construction. Even though it retained the same engine and transmission as the last of the XK 150S models, it would become the sensation that the XK 120 had been over a dozen years previously. So advanced was the XK engine at its

More Comfort, More Power – The XK 150

introduction in the late 1940s, that it could still manage to power the fastest standard production car in the world in 1961 and would not really start to be considered dated for another ten years after this. Altogether, this remarkable power unit would power brand new Jaguars in six different decades. The XK engine was not only the heart and soul of the XK sports car, but also of the company.

XK 150 3.8S Fixed-Head Coupé (1960)

Engine
Type	DOHC, 6 cylinders, 7 main bearings
Material	Chrome iron block, aluminium head
Dimensions	87mm x 106mm, 3,781cc
Induction	Triple 2in SU carburettors, 'straight-port' head
Maximum power	265bhp SAE (185bhp DIN) at 5,500rpm
Maximum torque	260lb/ft (gross) at 4,000rpm

Transmission
Clutch	10in single dry plate, hydraulic operation
Gearbox	4-speed close ratio, synchromesh gearbox (not first gear), with overdrive
Final drive	Hypoid bevel, ratio 4.09:1

Suspension
Front	Independent by torsion bars, double wishbones, telescopic shock absorbers and anti-roll bar
Rear	Live axle, half-elliptic springs and telescopic shock absorbers

Steering — Rack and pinion

Brakes — 12in discs, with 6.875in vacuum servo

Dimensions
Length	177in (4,495mm)
Width	64.5in (1,640mm)
Height	55in (1,395mm)
Wheelbase	102in (2,590mm)
Track – front	51.5in (1,310mm)
Track – rear	51.5in (1,310mm)
Turning circle	33.1ft (10.1m)
Frontal area	18.2sq ft (1.69m^3)
Coefficient of drag	0.47 (approx)
Weight	3,250lb (1,475kg)
Distribution	50.5/49.5

6 Racing – 1949 and 1950

The extraordinary top speed attained in front of the press by HKV 500 in May 1949 would silence the cynics who said the car could never reach its claimed speed, but there still lingered those who carped that the car might be fast in a straight line, but would be no good on a race track. In a way, Jaguar invited these comments by offering a car with unrivalled performance at about half the going price. Nevertheless, when the *Daily Express* announced that it would include a one hour race for production cars in its International Trophy meeting (held jointly with the British Racing Drivers' Club) at Silverstone in August, Lyons saw the obvious opportunity.

RACE TESTING

Fast as the XK may have indeed proved in a straight line, Lyons did not wish to enter

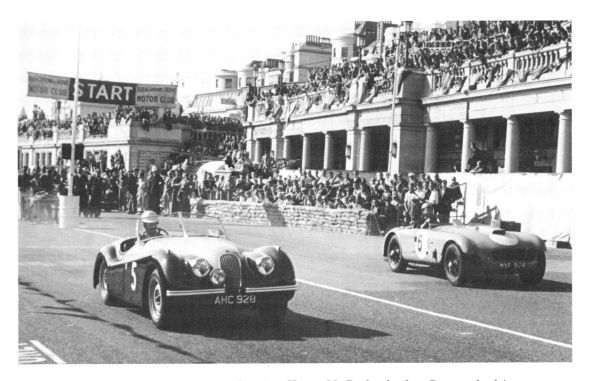

In its day, the XK 120 had supercar acceleration. Here a Ms Parker leads a Connaught driven by a Ms Sarginson off the line in the Brighton Speed Trials.

this race unless he knew the XK could win. To this end, in July 1949 one of the three prototypes was flogged around Silverstone in secret for some three hours, driven by Wally Hassan and Lofty England. It is recorded that Lyons also arrived for a quick turn in the car, despite having forgotten to bring his spectacles.

Back at the factory, Hassan prepared a report of some adjustments that were needed to improve the car's behaviour in a race – this would represent the beginnings of the Competitions Department, although this may not have been fully appreciated at the time. Hardly surprising when viewed retrospectively, a great number of the points requiring attention concerned the brakes. Other points that needed correction were the water temperature, gear ratios, tyres and a problem with wheelspin. These are all fairly routine problems that a standard condition road car would face when being race-prepared. The position of the pedals and the lack of support from the seat – no belts to hold you in place in 1949 – were also highlighted.

DRIVERS SOUGHT

It would not be much use carrying out such careful race-preparation on even the fastest cars without employing top-flight people to drive them. In this respect Jaguar were just as canny, for they managed to obtain the services of three excellent drivers for the three prototype XK 120s, one of the drivers being just about the fastest available.

Prince Birabongse of Siam was at the top of his form. Apart from the Production Car Race, he was also entered on his 1,500cc supercharged Maserati – drivers were described as sitting 'on' their single-seaters in those days – for the day's main race, the International Trophy, for Grand Prix cars. There was not quite yet such a thing as World Championship Grand Prix: these would begin at Silverstone the following May. In this first heat, he beat Ascari – a man who would hold the distinction of winning no fewer than nine Grand Prix races consecutively. Bira had a connection with Jaguar, as Lofty England had worked as his racing mechanic while employed by Bira's cousin Prince Chula who ran the White Mouse racing team before the war.

Leslie Johnson had raced in just about every category with some success. He had been highly placed in rallies, hill climbs and Grand Prix races, as well as in sports car races, where he had co-driven the winning car in the 24-hour event at Spa-Francorchamps, with Jock St John Horsfall. At the time he was generally considered to have just moved into the top rank of drivers. His connection with Jaguar was a personal friendship with Bill Heynes, and it was in fact Johnson's BMW 328 that had had its engine dissected when Jaguar were experimenting with power units before the 'XK' was settled upon.

Peter Walker completed this trio and was chiefly renowned at that time for his extremely spirited driving of an ERA, pre-war. His occupation was farming in Herefordshire, yet despite the amateur status of his race driving, Lofty England apparently reckoned that Walker could out-run Moss if he was in the right mood. This is some claim. He would go on to win Jaguar's first Le Mans with Peter Whitehead in 1951.

ONE HOUR RACE

The One Hour Production Car Race on 20 August 1949 was seen by Britain's motor makers as a great opportunity to show off

Racing – 1949 and 1950

Getting ready for their first race, the three XK 120 prototypes. Is that William Boddy of Motor Sport *on the left?*

their cars' turn of speed. The organizers had, unfortunately, turned down the factory entries of Aston Martin and Connaught, but Allard, Healey, Riley, Frazer Nash, HRG (Halford, Robins and Godfrey – the firm's three partners), Jowett, MG, Morgan and Jaguar all sent their works (or quasi-works) cars. There were also private entries, among them a 1936 Aston Martin, two rather large Lagonda four-seater Rapides and an SS 100. This was without question the most exciting production car race yet seen in Great Britain. The XK 120s certainly had plenty of opposition – on paper.

Bira took pole position in 2m 10s, a time that would have put him on the grid for the Grand Prix race. (Progress is such in motor racing that it is chastening to note that in the relatively new motor sport of truck racing, the huge six-ton tractor units have posted faster laps times around Silverstone than Bira's in the XK.) Peter Walker and Norman Culpan shared the second fastest practice lap with 2m 11s. Leslie Johnson was joint sixth fastest, alongside an Allard and two Healeys. Their time was five seconds slower than pole.

Bira's car was the first XK of all, HKV 455, chassis number 660001. It had been repainted blue in his racing colours, with yellow wheels and steering wheel! Walker would drive 670001, a white car painted red for the race and Johnson had HKV 500, chassis 670002, the Jabbeke record car and still white. Both Walker and Johnson's cars were converted from left-hand drive for the race, because Silverstone's preponderance of right-hand corners – as with most motor-racing circuits – showed up a lack of traction from the right-hand rear wheel. This was most obvious in the lower gears and moving the driver's weight across to the right would help to push down the offending wheel on to the track. Future Jaguar two-seater competition cars would all be right-hand drive.

Racing – 1949 and 1950

On the grid. The strange sidelights on the race-winning car, HKV 500, can clearly be seen to be smaller than those on the second-place car behind it.

It was a Le Mans start – oh, how these are missed! – and it was an Allard that was first away. Potter's K-type Alpine Trial model with almost one litre more than the three Jaguars charged off with Johnson (number seven) and Walker (eight) right behind, with Bira's number six a couple of cars down in the rapidly accelerating jam. At the end of the first lap the first three cars were Jaguars, with Johnson still leading Walker and Bira. Amid the flurry of squealing tyres as the confusion of metal hurtled over the finish line in pursuit of the three XKs, Lyons and the Jaguar team could not have helped feeling a great surge of pride. This would be tempered, however, since there still remained almost an hour of racing.

There was a fair distance already between the three Jaguar roadsters and fourth place, this being Tony Rolt in his Healey. He was followed by Norman Culpan in a Frazer Nash and Len Potter's Allard which has led at the start but was now down to sixth spot. Very soon Culpan would get by Rolt and set off after Walker who was now in third place, some distance behind his team mates. By lap four, Bira had found a way past Johnson and all three Jaguars were beginning to lap the back markers, one of whom 'lost' the rear end of his Jowett Javelin, causing Johnson to spin into the straw bales serving as an Armco barrier. He resumed soon enough, but the nudge with the Jowett had left the nearside front of HKV 500 no longer as originally designed.

Racing – 1949 and 1950

The start. Len Potter in his Allard has made the sort of start that became the speciality of Stirling Moss, although he would not be ahead for long.

This meant Culpan's Isleworth car, with its six cylinder BMW 328 engine, was now in third place and beginning to snap at the heels of Walker in the red XK. Johnson soon caught up these two, but he now had great difficulty in ridding himself of the very well driven Frazer Nash, which actually got by him on lap nine. Gradually, however, Johnson began to pull slightly away from Culpan, passing Walker into the bargain. After a few laps, Culpan managed to do the same and the leading four were now Bira, Johnson, Culpan and Walker.

The crowd, enjoying some true halcyon mid-summer weather, were, of course, loving all this. The spectators had already read about the remarkable top speed capabilities of these new Jaguars and most of them were now seeing the cars for the first time. (There were still only about half a dozen 120s made.) As the Jaguars streaked on ahead of the field, the spectators could now confirm the cars' superb lines, while being treated to a demonstration that the XK 120 was properly as fast as it promised – not forgetting Culpan's Frazer Nash. They also had much to entertain them in the cars that the Jaguars outpaced that day.

Down the field there were many spills as several competitors caught those wretched straw bales, fortunately without injury in this race. (In the International Trophy final, St John Horsfall would, sadly, lose his life when his ERA inverted after contact with a bale. Not only was he driving

Racing – 1949 and 1950

Leslie Johnson pitching HKV 500 hard at a corner, trying to make up for time lost spent denting the front wing.

the same car that ended John Bolster's motor racing days earlier in the year, but he crashed at the same corner as Bolster.) There were slides, gyrations, sick-sounding cars and complete breakdowns, as well as the odd collision between competitors. It is surprising that there was relatively little competitor-contact in the race, for a great deal of baulking went on, thanks to the great speed differential between the fastest and slowest cars. For sure, the crowd were having a whale of a time: most reports afterwards claimed it was the race of the day. However, the greatest drama of the race had yet to occur.

Bira had been circulating rapidly at the front of the race and by half distance, at 14 laps, he was five seconds ahead of Johnson who was about 1½ seconds ahead of the inspired Culpan who, in turn, now had Walker right on his tail. The lap speeds for this bunch were improving all the time and when Walker found a way past the Frazer Nash, Johnson seemed to find a little more too, for he was now reeling in the Siamese prince. Just as he was a car's length behind, with the two of them sweeping into the infamous Woodcote Corner, Bira's nearside rear tyre burst, sending the race leader into the bales. Apparently, Johnson very nearly went with him for his second trip into the straw, but just squeaked past the sliding blue XK, while Walker had a bit of a 'moment' in avoidance as well.

The race continued apace to its conclusion, which occurred as soon as the leader, Johnson, crossed the line after the hour had elapsed. Walker followed him five seconds behind, with Culpan coming home third, fourteen seconds behind the red XK.

Had Bira finished, Jaguar would have won the team prize, but failure to capture this never really matters too much to those who win.

Bira had actually made a brave attempt to change the deflated tyre and wheel – which led to urgent calls from the loudspeakers for an official observer to oversee his actions. Alas, the jack dug into the grass and he was unable to get the necessary clearance to remove the wheel. It seems that if he had motored gently forwards just a few yards, he would have gained some hard standing, could have changed the wheel and reached the finish, albeit a lap or two down on the winner.

The reason for the sudden tyre failure seems to have been caused by heat build-up on the almost constantly loaded left-hand rear tyre from Silverstone's repetitive right turns. Not only was it a very hot day, but the Jaguars were almost cheekily still wearing their rear spats. Coupled with this, it was found after the race that this particular wheel on 660001 had been fitted with a touring inner tube inside its 6.00 x 16 Dunlop racing cover. It is also reported that the tyre was rubbing on the bodywork inside the wheel arch, adding further to a rise in the tyre's temperature that would deny a Jaguar XK 1-2-3 in the model's first track appearance. There was no reported problem of tyre-fouling bodywork on the other two cars, but Bira's car was the first XK of all and was most certainly different from the other two, even if they were prototypes as well.

There is a misleading notion that had all Bira's tyres retained their air, the XKs would have finished at the front in a romantically patriotic red, white and blue sequence. This would have been white, blue and red, surely, since Johnson had shown after his spin that he could catch up his team mates relatively easily and was about to take the lead from Bira when the tyre blew. Unless, of course, he could only reel them in because Bira and Walker were holding back very slightly to conserve their brakes (even though special drums had been fitted for the race) and were going in for a sprint finish in the last couple of laps. Since these three drivers are, sadly, no longer alive, we may never know if any such tactics were planned.

THE DREAM MACHINE AND ITS RIVALS

Nevertheless, despite the mishap of the blue car, the XK Jaguar had now established itself as just about everyone's dream machine. Jaguar must have been very pleased with the outcome, although shrewder members of the company would have realized that Silverstone was not particularly hard on brakes and that the advantage the XK had in its engine performance was not held in all other areas of the car. The showing of the former racing motor-cyclist Culpan in his 2 litre Frazer Nash must have confirmed these doubts for some people.

His Frazer Nash had cornered with very little roll and he had made the fastest lap of the race, despite giving away nearly 1½ litres in engine capacity. Indeed, looking only at photographs of the race, anyone not knowing the result might be surprised that the XKs had finished at the front. Not only did they look far more standard than most of the other cars – their hub caps and rear wheel covers were in place for a start – but the roll angles they displayed are alarming. Truly they looked like standard touring cars being raced merely for fun. Apparently, they were a lot quieter than most of the other cars as well, just to add to their 'touring' image. It is quite extraordi-

Racing – 1949 and 1950

'Where's Culpan?' No doubt Peter Walker is wondering whether he's succeeded in shaking off the 2 litre Frazer Nash.

nary that this race track superiority had been achieved in just a dozen years of peace-time production, from the first SS cars in 1931 to the first deliveries of an XK-engined Jaguar in 1949.

Any disquiet the factory may have had about the Frazer Nash as a rival on the racetrack cannot have extended to unease over the car as a rival in the showroom stakes. The Le Mans Replica Frazer Nash may have been almost as rapid as an XK 120, but it was really very stark by comparison, with little of the Jaguar's interior appointments and refinement, and almost none of its ride comfort. Furthermore, it had all the elegance of a frog.

Jaguar's dominance over the other makes in the race would prove a portent of its performance in the car industry over the next few years. Of the eleven different makes of car in this race only three others remain: Aston Martin, MG and Morgan. At the time of writing MG had only just offered a sports car for sale (the MGF) after many years, and the only Aston with a realistic chance of surviving is a re-bodied Jaguar. It is also a fair bet that it was the XKs and Jaguar saloons that brought about the demise of Allard, Frazer Nash, HRG and Jowett, while Lagonda appears also to be currently extinct.

MORE RACES IN 1950

Back at the factory, plans were made to enter some more prestigious races with the 120. When it was announced in the press that the Royal Automobile Club's Tourist Trophy was to be revived in 1950, Hassan was quick to produce a lengthy report that outlined the XK's competition shortcomings, coupled with what was little more than a warning that entry into this, and any other, race, must be taken very, very seriously. Before this RAC TT race in September, where an XK 120 would make a star overnight of one of the finest drivers in any type of car the world has seen, the XK would compete in America, the Mille Miglia, the Alpine Rally and the already popular Silverstone Production Car race.

With British interest in the XK 120 now at a positively rabid level, Lyons hoped to generate similar enthusiasm in America by means of racing the car out there. To this end, at the end of 1949, Leslie Johnson found himself on his way across the Atlantic to take part in a race at Palm Beach in Florida, run by the Sports Car Club of America.

In this, the first race outside Britain for the XK, Johnson held second place for many laps behind a frightful hybrid of a successful Indianapolis racer and a modern Ford V8 engine. Called a Duesenburg-Ford, it was in truth a far faster car than Johnson's – which was the red Silverstone car, 670001, but still right-hand drive. After a compulsory pit stop, Johnson dropped to fourth place where he remained behind this Ford Special, the Healey of Briggs Cunningham with its large Cadillac V8 motor, and a 12-cylinder Ferrari 166. Of two other XKs entered, one gained eighth position and the other ran out of brakes. Being a street circuit, the brakes would have been called upon to work much more frequently than at Silverstone. Johnson himself reported at the finish that his brakes could have done with improving. This may be a good example of British understatement.

Although he did not win the race, Johnson and Jaguar at least found some satisfaction in the acknowledgement of the standard nature of his mount, compared to those who finished ahead of him, for he won the trophy for 'best production car' presented, ironically, by Donald Healey. Further publicity was gained for Jaguar in North America by the use of an XK as the official course car. The XK would not wait long for its first foreign win, for the following month in Cuba, Alfonso Mena beat a motley assortment of machines in a production car race.

After the Palm Beach round-the-houses race, the red Johnson car was sold to Jack Rutherford who would also compete with it. Most happily, this prototype 120, the first left-hand drive one and second XK to be made, now resides in the Walter Hill collection – it is the only surviving one of the three prototypes. It remains right-hand drive and has been repainted its original white. The frequency with which these first three prototype 120s changed colour really does make you wonder whether their identities ever became muddled.

Back in England, the factory had decided to step up its competition efforts with the XK 120 and six cars were prepared in the spring for selected drivers to race. These were aluminium-bodied cars, all right-hand drive, with chassis numbers 660040 to 660044 and 660057. Five were for road and track racing, of which four were registered for the road, bearing the letters 'JWK'. The sixth car was earmarked for rallying. They were all officially privately entered cars, but considering they were provided and backed by the factory, to deny that they were really works cars is something of an exercise in semantics.

No. 660040 was allocated to Leslie Johnson and was registered JWK 651; 660041 went to Nick Haines and seems to have been unregistered. It is interesting that cars which did not have the easy

Racing – 1949 and 1950

identification of a registration number have become far less well-known than the HKVs and JWKs of Jaguar folklore. No doubt, they would eventually be worth less, although that is hardly of much significance here.

The third car, 660042, was prepared for Peter Walker and bore the registration JWK 977, while 660043 went to the Italian Clemente Biondetti. Registered JWK 650, it was the only one of the six that remained in Jaguar's ownership. No. 660044 was prepared for Ian Appleyard, who would achieve great success with NUB 120 in rallies. (This car is usually considered the most famous 120 of all, although HKV 500 should really carry this accolade.) No. 660057 was allocated to Tommy Wisdom and bore the number plate JWK 988. This was the penultimate aluminium right-hand drive car to be produced.

It was actually 660043 that was the first car to be completed, in time for the Targa Florio. This was one of the oldest motor races in the calendar, having started in 1906. It may also have been one of the most dangerous. Clemente Biondetti was a driver whose previous achievements in this event and the Mille Miglia were unique. He won the race in 1938, and again in 1947, 1948 and 1949 – the only person to achieve the hat-trick in this event. Also in 1948 and 1949 he won the Targa Florio, meaning a win in 1950 would produce the first-ever hat-trick in the Sicilian race as well. Lyons was certainly continuing the policy of picking the best drivers for the job.

Biondetti had already had a taste of XK driving when he and HKV 455 were introduced, after it was taken across to the Geneva Motor Show in March 1950. Lofty England would be the one who would

Biondetti's car, being raced with great success some fifteen years later by Rhoddy Harvey-Bailey.

conduct the XK on this mission – to think he would drive the world's fastest production car on such a trip and get paid for it into the bargain!

As the Targa Florio got under way, Biondetti in his Jaguar was holding second place ahead of several V12 Ferraris. These were driven by such men as Luigi Villoresi, who had won the Targa Florio twice and who would win the Mille Miglia the following year; Giovanni Bracco, who would be second to Villoresi in the 1951 Mille Miglia and winner himself the year after; and Giannino Marzotto, who would win the Mille Miglia that year and in 1953. Biondetti and JWK 650 were leading some of the most successful road racing drivers of the day.

The only driver ahead of Biondetti was Alberto Ascari in his V12 Ferrari. Possibly, it was only Ascari's sheer virtuosity that was keeping his Ferrari ahead of the XK. The Italian would, after all, be runner-up in the Formula One World Championship that year and champion in 1952 and 1953. Indeed, in 1952 he would win every Grand Prix that he entered, albeit in the absence of Fangio. Such skill allowed Ascari to extend his lead slightly, although over a race distance of more than 600 miles (1,000km) anything can happen. Perhaps it did.

The Biondetti hat-trick evaporated with a loud bang, for as half distance approached JWK 650's number three connecting rod elected to ventilate the cylinder block and Biondetti and his passenger – not a co-driver – coasted to a halt, temporarily enshrouded in oily smoke. The rod had suffered a failure just above the big end bearing as a result of a minute flaw in the forging.

It will not have been much comfort for Biondetti, who never won this race again, but this failure would mean that all Jaguar connecting rods would in future be crack-tested before assembly. This is proof that 'racing improves the breed' indeed, because connecting rod failure would be very, very rare on a properly maintained Jaguar engine, not only throughout its competition life in the 1950s, but right up to the present day.

All the leading Ferraris themselves wilted under the rapid pace, including Ascari's. It is often written somewhat triumphantly that Biondetti caused them to fail because of the speed that his lone Jaguar had displayed in the race. Since the car from Coventry itself blew up, this is not much to be proud of. The race was won by the little-known Bornigia brothers, Mario and Franco, in an Alfa Romeo, ahead of two Ferraris piloted by drivers unknown outside Italian road racing. They were either simply not rapid enough drivers to strain their steeds to breaking point, or were skilful enough to bring their Ferraris home where others had failed.

THE 1950 MILLE MIGLIA

Biondetti had some interesting comments about his XK 120 after the 1950 Targa Florio. It seems he loved the performance, liked the steering, thought the front suspension was satisfactory, was not sure about the rear suspension, and said the brakes were poor. He was a good judge of a car, of course, and one would expect this from the man who had been Italy's best sports car driver, with four Mille Miglia wins to his credit. Even though he was born in the nineteenth century, Biondetti was still faster than nearly anyone at road racing in 1950 and he was eagerly awaiting the forthcoming Mille Miglia, just three weeks after the Sicilian race. Could he make it four wins consecutively in 660043?

Racing – 1949 and 1950

The fact of the matter is that the Jaguar XK 120 was a heavy touring car and the Ferraris, Maseratis and Alfas it would compete against were not only far lighter – the Ferraris were said to scale around 1,650lb (750kg) – but were really sports-racing cars of very limited production and of the most Spartan, even raucous kind. Nevertheless, Biondetti reckoned he could win this classic Italian road race again.

So did Jaguar, apparently, for they now entered – or helped the privateers to enter – four of the quasi-works XK 120s for the Italian classic. The drivers were to be Biondetti, in 660043, Leslie Johnson and Nick Haines in 660040 and 660041, and Tommy Wisdom in 660057. Wisdom was, like Biondetti, no spring chicken. Although he himself would not manage any wins in his car during the season, JWK 988 would establish the reputation of perhaps the fastest Jaguar driver of all time before the year was out. However, the 1950 Mille Miglia had to be endured first.

First of the XK 120s to hit trouble would, alas, be Biondetti's. Not long into the race,

Some owners who competed in their XKs were obviously unconcerned about extra weight.

the car begin to misfire, but this was rectified by a change of spark plugs. No. 660043 then began to overheat and this necessitated regular topping up at every stop. As if this were not enough, a leaf spring then snapped.

The Mille Miglia was a notoriously bumpy ride, with all manner of pot-holes, ridges and even stretches that would be classed as 'unmade road' today. Another hazard would be very abrupt ramps onto the many makeshift bridges that were standing in until the originals could be properly rebuilt after the wartime destruction. That none of the other XKs broke springs, despite covering greater distances than Biondetti, could be explained by the fact that same rear springs had been on his car from the Targa Florio, another chassis-pounding event.

Despite the cracked spring, Biondetti was still determined to continue and motored on at reduced speed until he found a blacksmith who succeeded in strapping the spring together in under an hour. In spite of this bandaged spring and the water-hungry engine, he would still make second fastest time of the race on some stretches and managed to bring the XK home to the finish in eighth place. He reckoned that the time he had lost exceeded the 53 minutes that he finished behind the winner, Marzotto.

As for the other three Jaguars in the race, the Haines car flew into a ditch at three-quarter distance when co-driver Haller was at the wheel; retrieval took so long that they would be unplaced. Tommy Wisdom's effort lasted longer but the gearbox jammed fast, about 40 miles (65km) from the finish at Brescia. Leslie Johnson and Jaguar's John Lea had a virtually trouble-free run, apart from a minor electrical short-circuit, and brought JWK 651 home in fifth, beaten by three Ferraris and an Alfa driven by the greatest driver of them all, Juan Fangio. Considering the nature of these four cars when compared to Jaguar – the XKs had tried unsuccessfully to enter the race in the GT class – it was not a bad result, if a little disappointing.

JAGUAR/FERRARI HYBRID

Clemente Biondetti next raced 660043 in some minor Italian races, crashing heavily in the second and bending the chassis. The customs-induced delay while new parts were obtained caused him to make a sort of Ferrari-style Jaguar Special. This car was, Biondetti reckoned, some 1,100lb (500kg) lighter than an XK – but it was no longer an XK either. It was really a Ferrari 166 body and Biondetti-style frame, with Jaguar propulsion courtesy of the repaired Targa Florio engine.

It actually took part in the Italian Grand Prix at Monza in September 1950, but only managed 17 laps before retiring with engine trouble, having started twenty-fourth in a field of twenty-six cars. Uncompetitive it may have been but it was the only works Jaguar engine to take part in a Grand Prix. Biondetti would enter his Coventry/Modena hybrid for all manner of events, the car even being re-bodied to resemble a C-type for the 1952 Mille Miglia. Sadly, by this time, this remarkable man, who strove so hard to put his Jaguar ahead of his country's own products, was far from well and would, alas, succumb in 1955.

THE 1950 LE MANS

The next project for Jaguar and its new team of alloy XK 120s was this 24-hour race for sports cars and prototypes, held at the Sarthe circuit in France. Three cars were to be entered, 660040, 660041 and 660042. The pairings were Leslie Johnson and Bert Hadley, Nick Haines and Peter Clark, and Peter Whitehead (Peter Walker being unavailable) and John Marshall. No. 660057, the Wisdom car, was not entered for the French classic (Wisdom was already contracted to drive a Jowett) but went off to a race in northern Portugal a few days before the 1950 Le Mans.

Wisdom managed to start the race in Oporto in second spot and he maintained this position behind an Alfa Romeo with a 1,000cc advantage over the XK. Unfortunately, despite being fitted with Al-Fin drums, his brakes had worn out in the closing stages and he dropped back a place. After the race, JWK 988 graced the showroom of the local Jaguar distributor for just one day and attracted an enormous amount of interest.

Taking place every June, the race at Le Mans was, for prestige purposes, The Big One. Lyons fully appreciated this, naturally, but the 1950 French 24-hour race was really a toe in the water for Jaguar. Although carefully prepared, the cars were really quite standard. Bumpers, rear spats, full windscreen and weather equipment were removed and such items as extra lamps, bucket seats and 60 per cent larger petrol tanks were fitted. As tools and certain spares could be carried in the cars, their overall weight would be no lower than a 'stock' model; indeed, with the 24-gallon tank full, they were probably a good 3,000lb (1,375kg) as they awaited their drivers at the traditional 4pm start. It would subsequently be claimed by Jaguar that the three cars' engines were absolutely undeveloped, producing no more horsepower than those found in any XK sold to the public. This assertion surely displayed a refreshingly open-minded attitude as to fact.

As the race settled down, the Whitehead and Johnson cars lay in sixth and seventh places with two Ferraris, two Talbot-Lagos and an Allard (much improved since Silverstone the previous year, it seems) ahead of them. The Haines car was driven a little more gently in twelfth place. However, the Whitehead/Marshall Jaguar soon fell back as far as nineteenth with brake trouble and in fact made it to the end in fifteenth place. The Haines/Clark XK managed to retain its twelfth spot at the finish, although they had got up to eighth at one point. They fell back with clutch slip, apparently because of an oil leak. Perhaps, however, brake problems had been causing the gearbox and clutch to be overworked to slow the car?

Meanwhile, Johnson and Hadley circulated JWK 651 rapidly, picking up places as others retired. At the 15 hour mark, this brought them into second place behind the Talbot Lago of Mairesse/Meyrat which had inherited the lead when a faster team mate had been delayed in the pits. Not only was the Johnson/Hadley Jaguar on the same lap as this leading car, but they were gaining on it, for the Talbot was a little under the weather and would not win.

Alas, nor would Johnson's Jaguar, for with around three hours of racing left, the XK's clutch failed. It was not a case of the clutch slipping, for some kind of temporary fix can usually be brought about on even a chronically weakened clutch to allow a car to limp to the finish of a race. The friction disc had actually broken in two, with the splined centre that drives the gearbox input shaft parting around its cushioning springs and separating from the friction surface that is gripped ('clutched') by the pressure plate bolted to the flywheel. Thus was halted any drive to the rear wheels which could only be restored by removal of the gearbox and a change of clutch. To be classed as a finisher at Le Mans a car must be driving at the end of the 24 hours, but the breakage had happened away from the pits, and as no centre-plates were among the spares carried in the car anyway, that was that.

It would be wrong to think that at least the car's brakes held out (an XK 120's brakes lasting 24 hours at racing speeds?), for they had been quite ineffective for several hours and much changing down through the gearbox to slow the car for the corners had been necessary to keep in second place. The reverse loadings that this sort of driving produces put a great strain on the gearbox. That, of course, means a huge strain on the clutch, and with three hours to go, the clutch plate on Johnson's car had met its breaking point. It was therefore the brakes that had caused the car's retirement, just as they had caused the other two XKs to drive more slowly and finish further down the field.

The winning car was the Talbot-Lago driven by Rosier, father and son. This was followed home by the other Talbot of Mairesse/Meyrat and in third place – an excellent result – was the Allard of Sidney Allard and Tom Cole. It is always written that the Talbots were thinly disguised Grand Prix cars and thus not really fair contestants. It is certainly true that the two leading cars home were virtually the same as the Grand Prix versions, but with wings and headlamps added.

However, the origins of the 4½ litre Talbot Lago car lie in a two-seater version that raced pre-war. This was then developed into the Grand Prix car, and a not very good one at that. Thus the winning Talbot Lago was fundamentally a mid-1930s sports car. This said, with its engine producing over 200bhp to propel a weight of around three-quarters of a ton, it is all the more impressive that the leading Jaguar, even with its superior top speed, managed to get within striking distance.

It was rather unflattering, however, for the other two XKs that they were preceded home by two 5.4 litre V8 Cadillacs. This was not a make of car noted for its prowess on race tracks, and one was a virtually standard saloon, a 'Sixty-one' Series Coupe de Ville with enough chrome on the front to make about two dozen XK grilles. As ostentatious as it may have looked, a stock 1950 Cadillac would apparently out-drag an XK 120 up to 90mph (145km/h). The other was a great lump of a thing, being essentially another Coupe de Ville with a special all-enveloping body sitting on it. Perhaps Jaguar were mollified a little bit by the fact the French decided to call it 'Le Monstre'. That the Cadillacs were entered by Briggs Cunningham may have something to do with their very impressive French outing.

As it turned out, the race was an absolute celebration for British cars. Only two of the thirty-one cars that failed to last the twenty four hours were British: the Johnson Jaguar that had held second place, and an Aston Martin that was a last-minute stand-in when one of the team's cars was wrecked on its way to Le Mans. Indeed, the top twenty finishers contained no fewer than fourteen cars entered by nine different British manufacturers: one Allard (also with a 5.4 litre Cadillac power unit), two each of Healey, Bentley, Aston-Martin, Frazer Nash and Jaguar, followed in by one Jowett, one Riley and one MG. All five favoured Ferraris failed to finish.

The broken clutch plate that lost the Johnson/Hadley car third, or even second, place at the 1950 Le Mans would at least provide a useful lesson. As with the connecting rod that broke in the Mille Miglia, this failure brought about a swift and permanent change. In future, all Jaguars prepared by the factory for racing – where

Racing – 1949 and 1950

comfort is of far lesser importance – would dispense with the cushioning springs and have solid centres on their clutch driven plates.

THE 1950 SILVERSTONE INTERNATIONAL TROPHY

After the usual debriefing back at the factory, plans were soon being hatched for a more specialized car to enter Le Mans in 1951 and take on the likes of the Talbots. The next race for Jaguar and their sports car was the production car race at the Silverstone International Trophy, the same meeting at which the XK made its competition debut in 1949. Hopes of a repeat of the previous year's emphatic victory would be tempered somewhat by the presence of some foreign cars, notably the Ferraris.

With three of the cars repaired after their gruelling French race, as well as a general re-fettling of Wisdom's JWK 988, all four were given lower axle ratios, larger diameter torsion bars and ventilated brake drums. HKV 500 would also receive this treatment, made ready at the last minute for Adolfo Schwelm, a driver from Argentina. The car was also painted red for the race. (The spray gun had been working overtime on this car: it had now been bronze, white and red.)

Come race day, Johnson, Walker and Wisdom would be in their usual cars, JWK 651, JWK 977 and JWK 988. The unregistered Haines car would be driven by Tony Rolt, while HKV 500's driver was the subject of a sort of musical chairs. Tazio Nuvolari, claimed quite reasonably by

Rare shot of the amazing Tazio Nuvolari practising for the Production Car Race at the International Trophy meeting at Silverstone, 1950. Very regrettably, he was too ill to drive in the race.

many to be the greatest driver of all time, had arrived at the race meeting under the impression that he was to drive a Healey Silverstone. However, J. Duncan Hamilton would correctly prove the Healey was his for the race. Fearing great spectator disappointment, the British Racing Drivers' Club approached Lyons who saw the publicity opportunity in a flash and offered the Italian maestro a drive in HKV 500. Alas, after a rather uninspiring practice, Nuvolari, nearly sixty years old, was too ill for the race and the car was taken over by Peter Whitehead. What Adolfo Schwelm thought of losing his drive is not recorded.

HKV 500 did not get too far in the race after all this fuss, for an oil plug fell out (somebody must have sported an awfully red face for this) and Whitehead found that his engine ceased to operate rather suddenly. Poor Leslie Johnson, the winner of this race the year before, spun on this oil slick while putting on a stirring drive after his car had spent rather a long time deciding whether to start at flag-fall. He would recover to finish eighth, right behind Tommy Wisdom. Walker and Rolt had streaked ahead at the start and would finish almost effortlessly in that order.

However, Jaguars did not actually win the Production Car Race at the 1950 International Trophy meeting. The race was run in two heats, the first for cars of 2,000cc and below, and the second for those from 2,001 to 3,000cc and 3,001cc and above. There was a further (and highly relevant) distinction in that the first heat was run in the dry and the second on a damp track, it having rained in between. This meant that Ascari's Ferrari, which won the first heat did so at a faster pace – less than 2mph (3km/h) – over the one hour race than Walker's Jaguar, which won the second. Walker s speed in winning his heat in 1950 was almost exactly the same as his speed in coming second the previous year, despite the benefit of his car now having stiffer suspension and improved brakes. Obviously a dry track would have seen an improvement in his speed. It is a shame that the two heats were not combined, but too many entries had been accepted for that. The XKs would probably have been slightly faster than the Ferraris at Silverstone in a one hour event at that time, but a driver of Alberto Ascari's calibre could possibly tip the balance in favour of the Italian car. The next big race would, however, yield an outright win for Jaguar, as well as firmly establish the reputation of a young man who would deservedly become a legend.

THE 1950 TOURIST TROPHY

Last held in 1938 at Donington Park, the 1950 Tourist Trophy was to be a three hour 'Sports Handicap' – a handicap race for sports cars – and would be held at Dundrod, west of Belfast, in September. This new circuit in the hills was very tortuous with just one short straight. It would seem to suit the agile car, hardly a fitting description for an XK 120. Four of the unofficial team cars had been entered. Tommy Wisdom, driving a works Jowett in the race, had also put an entry in for JWK 988. Haines and Rolt were down to drive 660041, Leslie Johnson was in his faithful JWK 651 and Peter Whitehead would race the much-used HKV 500.

Tommy Wisdom was then motoring correspondent for the *Daily Herald* and a family friend of Alfred and Pat Moss. Either their young son approached Wisdom, or was approached by Wisdom – depending whether you read Wisdom's account or Moss's – for the XK drive in the Ulster TT. Whichever way the offer was made, it was

Racing – 1949 and 1950

readily accepted. Moss was only twenty years old and had never handled a large car like an XK in a race, all his racing thus far having been confined to Coopers and HWM single-seaters. His first-ever victory was only two years previously in a 497cc Cooper in a hill-climb near Brighton.

It seems Jaguar themselves were not at first willing to give a 'works' drive to one so young and with no real experience of racing a large and powerful sports car. Should he fly off the road and hurt himself, or worse, the adverse publicity on account of this lack of experience would be disastrous. It seems Wisdom argued on Moss's behalf and they allowed the entry to go ahead, meanwhile lending Moss a blue 120 to use on public roads, to familiarize himself with such a car. This reticence over Moss's youth is reminiscent of the attempts by some parties to stop the 1964 World Title fight between Cassius Clay and Sonny Liston, in case the far younger man was badly hurt. True genius is often startlingly precocious.

After his first practice session Moss was 26 seconds behind the fastest driver, Leslie Johnson, who had managed a lap in 5m 39s. Moss was despondent as he climbed out of Wisdom's light green Jaguar XK 120, but was offered encouragement by Johnson. Perhaps very effectively, for the following day Moss was fastest with a lap in 5m 28s. Nick Haines was not so effective, bending his car so badly that he and Rolt withdrew. Whitehead, the current holder of the outright lap record at Dundrod in a Ferrari, was close behind Johnson.

As the drivers lined up for their gallop across to their cars they were already being lashed by the wind and rain. For once, Moss was beaten away and he followed Johnson around the first lap, getting by at the end of it. The conditions were atrocious and many cars went off, tents blew away and the spectators were soon quite bedraggled. Despite the conditions, the young man in his first sports car race came in

Johnson trying to keep up with Moss at Dundrod, 1950. He appears to be on the ragged edge in what were appalling weather conditions for drivers and spectators alike.

An ultimate XK? An apparently original 1950 steel-bodied XK 120 roadster, rebuilt to complete 3.8 litre 150S specification.

The feline grace of the rear haunches.

The inviting interior of an XK 120 fixed-head coupé, which combined 120mph performance with this much luxury for a reasonable price. No wonder so many other firms went out of business.

Fixed-head XK 120 displaying its graceful profile, an improvement in this respect on the roadster.

The shape of things to come. The third car home in the 1940 Mille Miglia. A BMW 328, a one-off design for that event.

A superb study of an XK 140 roadster, enjoying what it does best on a clear day: taking its owner for a good blast on the open road.

The bumpers of the XK 140 mask the bottom curve of the front wings, making the car look squarer, a view exaggerated by the windscreen of the drop-head coupé shown here. It is fun to see one being thrown about like this, however.

Compared to the XK 120, the XK 140 grille may have been far cheaper to make, but was also less attractive.

The instrument panel of the XK 150. By this time, chrome had found its way into the interior.

Crimson cruising. A fine shot of an XK and its owner who, commendably, has not gone down the 'wire wheels' path.

Many XKs are now used for enjoyment in classic rallies, such as this Milan-registered XK 120 with its two intrepid occupants.

This delightful car has never been restored because it has been well cared for by one owner for a long time. While being in far from concours condition, cars like this can be far more interesting than an over-restored example.

Beauty in its unspoiled form – an XK 120 roadster with disc wheels and spats. This was, after all, how William Lyons designed the car in the first place.

This XK 120 drop-head, chassis number 667006, was bought in 1966 for £65 and used as everyday transport for some seven years. Those were the days.

Beautiful cars about to be enjoyed in their natural habitat.

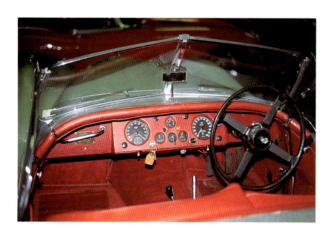

Revised instrument panel of a late XK 120 roadster.

XK 150 number plate and plinth. This is a late car with the larger rear lights incorporating amber indicators.

first, although not before his father in his pit had signalled him to go flat out on the last lap, after hearing that Bob Gerard had caught him up on handicap. He did this to such effect that he made the fastest lap, at 78mph (125km/h), on this last tour. He was to say in an interview in 1952 that he would not drive that fast in the wet again, for he 'now had more sense'. Whitehead finished second overall, after overtaking Johnson, who had slowed with brake trouble to finish third on the road, but only seventh on handicap.

Lofty England had nothing but praise for Moss after the race, despite his scepticism beforehand. He was impressed with the young man's self-discipline and the way he obeyed all instructions with great care. England also was delighted by the fact that when the brakes on Moss's car were checked over, they were found to be only half worn, where others had none left. It must be remembered that a wet race is kinder to brakes, since it is just not possible to use them so violently when adhesion between tyre and road is reduced. One wonders how long the brakes of Moss's team mates would have lasted on this twisty circuit had the weather been dry. As Stirling received his prize for this first important win, he had his twenty-first birthday to look forward to the next day: he must have been an awfully happy fellow just then!

SUCCESS ACROSS THE ATLANTIC

In America, a young driver had begun to win races in 1950 in his XK 120. After some good results in mid-season, notably second place behind an Allard at Santa Ana in California in June, which he might have won if he had not gyrated at the first turn, Phil Hill would progressively lighten the car and bore out the engine to 3.8 litres. Having been trained at the factory in England, he knew that this was the absolute maximum amount of metal that a 3.4 block could lose without collapsing. Entering the prestigious Pebble Beach Cup race, Hill yet again shot off course on the first corner in his first heat, again fighting back to come in second behind an Allard. He managed to win the final ahead of another XK 120 – presumably he had by now mastered the first bend of the race.

Phil Hill was talented enough to win the Formula One World Championship as well as Le Mans three times. He was abetted in his XK 120 adventures by Richie Ginther who would also become a Formula One Grand Prix winner, his single victory being just two fewer than Hill's total.

A look through the sports car race results on the other side of the Atlantic sees Allards beating the XKs on most occasions that they met on the tracks. Significantly for racing purposes, the Allards were usefully lighter than the XKs and had larger engines. Unfortunately, in the showroom stakes they were light years behind Jaguars, and actual car sales are ultimately of infinitely greater importance because a car company cannot survive to race without sufficient sales of its road cars. Allard sadly ceased producing cars in the 1960s.

Because such a large proportion of XK 120s were exported – home sales for 1949 to 1953 amounted to a mere 8 per cent – and because most of those that were exported went to America, much larger numbers of XKs were raced on the other side of the Atlantic. Apart from the soon-to-be-famous Phil Hill, Sherwood Johnston, Don Parkinson, John Fitch and Walt Hangsen all raced and won in XK 120s.

Racing – 1949 and 1950

Whoops! A fine view of an XK 120, courtesy of the errant driver in this American event.

Since the 120 in absolutely standard trim was too heavy properly to challenge the Allard and Frazer Nash entries – not to mention the Ferraris – many owners would have special lighter bodies made for their XKs, just as Biondetti did in Europe. There can be no doubt that the exploits of these, often amateur, racing drivers in their XK Jaguars would help start the ball rolling towards the great sales successes of all Jaguars in the United States in the 1950s and 1960s.

7 Racing – 1951 and Beyond

Jaguar must have been rather pleased with the win at Dundrod, for they signed Stirling Moss for 1951. Also, their XK 120 had at last won a second important race – albeit once again in the absence of Ferrari. Near misses – such as at Le Mans, the Mille Miglia and when the weather thwarted a second win in the Silverstone production car race – do not count; wins were what mattered and after two years, the XK had only two to its credit on this side of the Atlantic. Victories in rallies were fortunately coming a little more easily to the 120, as described in Chapter 8.

THE 1951 MILLE MIGLIA

One of the first races contested in 1951 was the Mille Miglia. Only one of the four semi-works drivers would be in his normal alloy-bodied car, however. Nick Haines apparently was not keen to enter any more of the arduous Continental road races, while Tommy Wisdom would be racing in an Aston Martin and Clemente Biondetti was still trying for his incredible fifth win in his Ferrari-Jaguar mongrel. This left Leslie Johnson in JWK 651 and he was joined by the new and deservedly hot property, Stirling Moss, who would enter in the old campaigner, HKV 500. It is now a motor racing legend that Moss would win the Mille Miglia with a record average speed that will never be broken. Alas, this was in 1955 and the 1951 race was not one of his, nor Leslie Johnson's, luckier ones.

Moss was accompanied by Frank Rainbow, while John Lea teamed up with Leslie Johnson as in the previous year's race. Moss drove over the course for four days before the start and several times he encountered Biondetti who on one occasion let him drive his special. Moss reckoned it was slower in top speed than the XK – cycle wings are an aerodynamic millstone – but far better in acceleration and cornering, thanks almost entirely to its considerable weight advantage. As for his own car, HKV 500, Moss found the brakes poor and the rear shock absorbers wrong. In an effort to stiffen the rear suspension for the notorious surface, both cars would have their rear springs bound tightly with tape, which Moss had fetched from Brescia. There was a third XK in the 1951 Mille Miglia, entered by a wealthy Covent Garden fruiterer named Gatti.

Johnson started at 4.29 am and Moss three minutes later. Despite the driving rain Moss took just 4 miles (6.5km) to catch the Ferrari that had started one minute ahead of him. Alas, this encouraging start would not last, for both Moss and Johnson would be out after covering about 15 miles (25km).

It seems that a 2 litre Fiat broke its engine and dropped a lot of oil at a particular corner and Ascari, reaching it in his Ferrari, had careered off, killing one spectator and injuring several. Soon after, Johnson would also crash there, his XK being too badly damaged to continue. Moss approached close behind to find someone waving him to stop. Neither braking nor twirling the wheel had any effect, for he also was on the dreaded oil and hit the small Fiat that had been parked across the escape road. It was no great consolation to Moss to discover that Ascari and Johnson had struck the same thoughtfully parked vehicle. No fewer than six pairs of competitors were consecutively eliminated at this corner, these all being at the very end of the starting order. The last competitor to leave Brescia was Clemente Biondetti and he was the only one of the last seven to get past this oily corner unscathed. This was partly thanks to the fact that Leslie Johnson ran back up the road to warn him and partly because he was Biondetti.

Meanwhile, Moss decided to struggle on to a garage to attempt to prise the wing off the wheel. But he overshot and his high-speed reversing manoeuvre caused the reverse gear wheel to attempt to friction-weld itself to its shaft. Freeing it after this perhaps unnecessary piece of 'flat-out parking' took two hours, but then the bonnet refused to stay shut, so he and Rainbow eventually called it a day. As for Gatti, he and his co-driver Fantuzzi also were early retirements when their XK 120 struck a road marker.

The Maserati-based chassis of Biondetti's special had proved so lacking in rigidity that, after fewer than 90 miles (150km) a high-speed 'yump' over a level crossing caused the fan momentarily to strike the bottom radiator hose. Or was this simply caused by the engine shifting momentarily on its mountings? Anyway, the resultant loss of coolant rapidly produced a fried engine and the end of Biondetti's 1951 Mille Miglia. Thus all four Jaguar-engined competitors were out well before half distance.

THE 1951 *DAILY EXPRESS* RACE

The next race was one where great things were expected of the XK 120s – the Production Car Race at the *Daily Express* International Trophy meeting at Silverstone. This would be the third running of this already very popular annual race, now held in May. As in the previous year, it was run in two parts, with cars under 2 litres racing separately. The factory's unofficial entry consisted of four cars, while several other privately entered XK 120s were also competing.

Walker was once again in JWK 977 and Wisdom likewise in JWK 988, but as HKV 500 could not be prepared in time, Stirling Moss would have the first left-hand drive steel car off the production line, 670185, painted light blue and registered JWK 675. This car was converted to right-hand drive, to help the traction on the many right-hand corners. It was found to be giving just under 200bhp on test before the race. Leslie Johnson's Mille Miglia mount being also unready, he would have KHP 30. When tested by Bill Boddy of *Motor Sport*, this car's brakes were found to fade. The only surprise here is that Boddy seems to have been the only journalist to have driven an XK 120 hard enough to find fade. Since road driving and race track work are poles apart, one has to wonder what was done to the brakes on this car to enable it to compete effectively at Silverstone.

In 1951, even though less than 15 per cent of XK 120 production was staying in Britain, sufficient quantities of the steel-bodied cars were now being produced to allow British drivers to start to acquire them and go racing and rallying. Of these privately entered XKs there were several, most notably Duncan Hamilton in his own car, LXF 731, making its competition

A marvellous photograph of the late J. Duncan Hamilton, looking a little like Mr Toad, at Silverstone's Copse Corner, 1951 International Trophy meeting.

début. Other XK drivers were George Wicken, Oscar Moore, Bill Holt and Neville Gee. Jack Broadhead had also entered an XK, to be driven not by himself, but by Allan Arnold. At the last minute Arnold was unable to race and a motor-cyclist named Charlie Dodson was asked to step in, despite the fact that he had not raced for twelve years. Non-Jaguar mounts in this second heat for the over 2 litre cars included the well-known names of Tony Rolt in his Nash-propelled Healey, Reg Parnell in an Aston Martin and the redoubtable Sidney Allard in one of his own.

In practice, Moss set the joint fastest time with Peter Walker. Moss reported being able to negotiate Woodcote at just below 100mph (160km/h) in one long slide. In the race, George Wicken took the lead, followed by Moss, Peter Walker and Duncan Hamilton. Wicken's lead was short-lived and on the third lap Moss took over, with Walker and Hamilton following him through past Wicken. One would have expected Walker to offer Moss some sort of a challenge, having been able to match the young star's pace in practice, but JWK 977 had developed throttle trouble and he was forced to make a pit stop, eventually finishing in fifteenth place. Moss moved into the lead on the third lap and drove away to win. Hamilton was pleased to bring his XK home in third place as he would claim his car was, at that time, still almost in standard trim. He was followed home by Wicken and Johnson. With an XK 120 taking second place as well, Jaguars would fill the first five places in this race.

While all the well known drivers were battling round the Northamptonshire airfield track, Dodson had moved stealthily up through the field, eventually finishing second. As Dodson on his 493cc Sunbeam had won his two Senior TT motor-cycle victories before Stirling Moss was born (one of them after falling off avoiding a spectator), it seems a fair bet that he was hardly a serious proposition before the race. However, his career as a top-class motor-cyclist might have brought to mind the extraordinary talents of Ascari, Nuvolari, Caracciola and Rosemeyer – all racing motor-cyclists who made it to the top in Grand Prix cars. Norman Culpan was another 'retired' racing motor-cyclist and his first-ever attempt at four-wheeled motor-sport gained him third place in the 1949 Le Mans. Motor-cycling tends to breed a superior sense of balance. Could this explain the great speed of William Lyons when he made the odd 'celebrity' appearance in one of his own four-wheeled products?

The impact of Dodson's excellent drive into second place – coming second behind Moss was really the same as winning an 'ordinary' race – was heightened by the fact that he never raced again. His drive was truly a one-off event: a fine way to bow out and create a bit of mystique in later years.

As in the previous year, however, the

George Wicken racing in the 1951 Silverstone International Trophy meeting in which he finished fourth behind Hamilton, having initially led the race. As with many privateers, the car would be used regularly on the road as well.

second heat saw some very fast times achieved by the under 2 litre cars. Both of these one hour Production Car Race heats were this year held on a dry track, and in the absence of Ferrari, a repeat of the 1949 result took place, with a wee Frazer Nash effectively coming third. A Le Mans Replica Frazer Nash won its heat just a smidgen slower than Dodson's time and therefore it covered a greater distance than all the other twenty-three finishers in the second over 2 litre heat, several XK 120s included. Roy Salvadori had led the field initially but inverted after tangling with one of those primitive bales of straw at Stowe. The winner was Anthony Crook of Bristol fame, and going by his average speed for the hour, his Frazer Nash would have finished about 30 metres behind the remarkable Dodson.

As in 1950, a really exciting race had been denied the spectators. Would it not have been more thrilling had the organizers, once again faced with a large entry, arranged the two heats by practice times and not engine capacity? Le Mans Replica Frazer Nashes, like the XKs, took the first five places in their race.

CLAES IN BELGIUM

In May in Belgium the jazz band leader Johnny Claes was lent an XK and was one of no fewer than eighty-four entries in the Belgian One Hour Production Car Race at Francorchamps. He won this race with ease, but there seems to be some doubt as to whether his mount was HKV 455 or HKV 500. Due to the Belgian plates, the car's identity is not obvious – another HKV riddle, perhaps?

Although his profession suggests no more than a superficial approach to motor racing, Claes contested some eleven Grand Prix races. Apparently the antics of private entrants of XK 120s were irritating the Belgian importer, the formidable Joska Bourgeois. This caused her to persuade Jaguar to lend Claes the car.

THE ARRIVAL OF THE C-TYPE

No sooner had the XK 120 become obtainable in Europe than the factory bought out the 'XK 120 Competition type'. This car

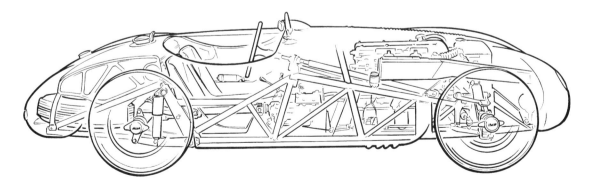

Never an XK 120, the C-type was built with the express purpose of winning the Le Mans 24-hour race.

Racing – 1951 and Beyond

was immediately dubbed the 'XK 120 C-type' by one weekly motoring magazine and later, of course, this would simply become the 'C-type'. It was no XK, however, although its designer, the aerodynamicist Malcolm Sayer who had recently joined the

---

XK 120 Type C (1951 Competition Model)

The car was dubbed 'C-type' by the press and appeared immediately before the 1951 Le Mans race for which it was specifically designed. In spite of its factory name, it could never be mistaken visually for an XK120, having quite different bodywork, chassis frame, steering and rear suspension. However, its engine, gearbox, rear axle, brakes (not 1953 models) and front suspension were developments of those items in the XK 120. Built for racing, weight-saving was a principal aim of the design, which achieved a reduction of some 770lb (350kg) over the touring car. As every Jaguar admirer will know, it won Le Mans in 1951 and 1953. Fifty-four were produced.

Engine
Type	DOHC, 6 cylinders, 7 main bearings
Material	Chrome iron block, aluminium head
Displacement	83mm x 106mm, 3,442cc
Induction	Twin 2in SU carburettors
Maximum power	210bhp at 6,000rpm

Transmission
Clutch	10in single dry plate
Gearbox	4-speed synchromesh gearbox (not first gear)
Final drive	Hypoid bevel, ratios to suit

Suspension
Front	Independent by torsion bars, double wishbones, telescopic shock absorbers and anti-roll bar
Rear	Live axle, transverse torsion bar, trailing links, right-hand A-bracket and telescopic shock absorbers

Steering   Rack and pinion

Brakes   Self-adjusting 12in drums – two leading shoe at front

Dimensions
Length	157in (3,990mm)
Width	64.5in (1,640mm)
Height	38.5in (980mm)
Wheelbase	96in (2,440mm)
Track – front	51in (1,295mm)
Track – rear	51in (1,295mm)
Frontal area	13.8sq ft (1.28m^2)
Weight	2,140lb (970kg)
Distribution	50/50

Racing – 1951 and Beyond

company from Bristol, did endow it with some hint of XK 120 in its bonnet line.

The C-type had a development of the XK engine and practically the same gearbox and rear axle, but with more suitable ratios. The front suspension and brakes were also essentially the same design as those on the XK, but developed for racing. On the other hand, the body, chassis (such as it was), steering and rear suspension were quite different. The most significant thing about the C-type was that it weighed 780lb (355kg) less than an XK 120. It would race unequivocally in the sports car category: no excuses this time, if beaten, that it was really only a touring car.

THE 1951 LE MANS

The C-type Jaguar was announced to the world only by its arrival for practice at Le Mans in 1951, a race that it won by nearly 80 miles (125km). With the three new Jaguars filling the first three places at the six hour mark, one began to think a minor miracle was occurring, but the two other C-types succumbed to engine failure.

At the 1951 Le Mans there was also a lone XK 120, registered AEN 546 and driven by two privateers, Lawrie and Waller. Considering the very limited competition experience of the former, their resultant eleventh place – the highest an XK 120 would manage in this race – was more than creditable since the opposition would be that much stronger than the year before and also because the 1951 race was frequently slowed by rain. Indeed, the 1,980 miles (3,186km) that they covered would have netted them the 1949 race.

After the arrival of the C-type, the major races would now be contested by this new car and so the XK would be used mainly by privateers without factory backing. Any thought that this meant the 120 would suddenly stop winning races can be quickly dismissed. For a start, total C-type

Now this is what you call trying! A helping of opposite lock is being served to stop the oversteer getting out of hand, at Prescott in 1951.

Racing – 1951 and Beyond

S.J. Boshier leading several XKs and an Allard at Goodwood in 1951. Strange today to see a Bugatti in this company, even if it apparently cannot keep up.

production would amount to less than 0.5 per cent of XK 120 production. This meant that thee were simply too few C-types around to enter all the races available, even had the factory wanted to enter them.

A postscript to the 1951 Le Mans is the existence of three lightweight XK 120 bodies, made for the race, in case the C-types were not ready in time. These were given the code names LT-1, LT-2 and LT-3 and the three magnesium bodies were carried on frames that in turn would be fixed to the XK chassis. It seems that these lightweight bodies did not actually find chassis until some time later, when a visiting American, the West Coast Jaguar distributor Charles Hornburg, espied them at the factory. He made a deal to buy two of them; these were put on to two chassis off the production line and taken to race across the Atlantic where he named them 'Silverstones'. Raced by Phil Hill and others, they achieved limited success. The third lightweight body, LT-1, would languish around the works for three years until being bought by a Jaguar employee.

THE 1951 WILLIAM LYONS TROPHY

It was a foregone conclusion that an XK 120 would win the William Lyons Trophy race at Boreham Wood in August 1951, as there were no other types of car racing. Duncan Hamilton had been hard at work on his XK (LXF 731) and he won the race without any real opposition. The fact that it rained may have helped this most colourful of all racing drivers, for he tended to excel in the wet. This is often a characteristic of the smooth driver.

As more and more owners now raced their newly acquired XK 120s, so the car became dated and gradually less competitive against more modern designs. Now that the C-type was the main Jaguar weapon on the race tracks, the XK 120 would often find itself making up the placings in races won by the C-type. However, many an owner would manage to bring an XK 120 to a level higher than the factory ever managed, now that they were concentrating development on the C-type. The XK 120 also won many club races in the absence of the C-type. Indeed, the XK 120 would continue to win club races for an extraordinarily long time after its introduction. It would clearly not be feasible to try to list all the races won by the XK 120 after the first couple of seasons were over, but one can at least remember some of the more famous car and driver combinations.

Michael Head losing the lead to Bill Holt, at the Easter Handicap at Goodwood in 1952. Holt appears to be waving his thanks as he speeds off to win.

Racing – 1951 and Beyond

John Swift leading a trio of XK 120s, also at the Easter Goodwood meeting in 1952. Wearing his customary collar and tie he would finish second behind Holt.

An immaculate XK 120 fixed-head coupé at Goodwood in 1954. The extra lamp is a good idea and its position does not detract from the car's appearance, but what about the possibility of overheating?

Racing – 1951 and Beyond

Some XKs raced were not so immaculate. This delightfully battered old thing leading a Lotus is seen on the Railway Straight at Aintree.

Towards the end of 1951, it was Hugh Howorth who owned the fastest racing XK in the country, with his (relative to the time) extensively modified car, registered as EBN 722, but seldom carrying any plates during races. Two readily recognizable features that this car would acquire were its two-tone colour scheme and a right-hand headlamp pod used as an air-feed to the carburettors. This latter modification was not unique, but the fact that it sported a mini XK 120 radiator grille across it certainly was. Many other XKs would go very well indeed, especially as their brakes improved a little with availability of wire wheels to help brake cooling, as well as improvements to the friction material. Howorth went on to acquire JWK 977, Whitehead's former car, onto which he grafted a factory-supplied C-type rear suspension. With several of his own modifications, including an engine bored out to 3.8 litres, this car with its black and silver colour scheme was the XK 120 to beat in the early 1950s.

Frequently competitive with Hugh Howorth was John Swift whose XK was now being extensively modified, if a limited slip differential was presumably not yet available.

109

Racing – 1951 and Beyond

THE 1952 'RACE OF CHAMPIONS'

Wins by XK 120s in 1952 were becoming harder to find in international races. The C-type won the Production Car Race at the International Trophy meeting – something of an XK 120 benefit in the three previous years – although a race at this meeting was arranged especially for the XK 120. This was called the 'Race of Champions' and as with the William Lyons Trophy race at Boreham the year before, all contestants were 120-mounted. This time, however, they were identical cars taken straight from the production line. It will therefore be no surprise at all to know that this race was won by the best driver in it, Stirling Moss.

S.J. Boshier having last-minute adjustments to his car. There are some intriguing cars behind.

Spinning at Goodwood in 1954. This car was seen for sale in Motor Sport in the late 1960s and it stood out because it was absolutely immaculate at a time when XK 120s were at their lowest ebb. Presumably it had survived undamaged during the interim.

THE 1952 LE MANS

High hopes centred on Le Mans for the C-type in June, but a total rout took place, with all three cars out within two hours. In an effort to gain some more speed down the Mulsanne straight, last-minute modifications were made to the shape of the cars. In consequence, these alterations had necessitated changes to the cooling systems and the additional pipes used to carry water to the remote header tanks were specified at too small a diameter. The main damage to the three engines occurred in practice and reversions to the original cooling systems on two of the cars came too late. Moss's car withdrew from lack of oil pressure, although this only can have preceded the terminal overheating problem by a few laps. The other two C-types were very soon out in clouds of steam, their cooling systems boiling merrily. The fact that the man who had to redesign the cooling system in a hurry was named Kettle would not have created much mirth at the Coventry factory in Browns Lane. Le Mans would provide several more dramas for Jaguars in the years to come, and one of these occurred in 1956 when another XK was entered, five years after the model's previous appearance in the race.

THE 1956 LE MANS

Two relatively inexperienced drivers, Peter Bolton and Robert Walshaw entered an XK 140 (PWT 846) in the 1956 French 24-hour classic. Not only that, but it was a fixed-head coupé, and a fixed-head XK 140 was

A gallant effort by this XK 140 in the 1956 Le Mans was thwarted by a rare infringement, no fault of the car itself.

the least sporty of the XKs at that time, simply because it was the most practical. Furthermore, this car had already covered 25,000 miles (40,000km) and although sent to the factory for race preparation a few days before the race, it certainly looked remarkably standard, even to the extent of retaining its walnut dashboard.

Apparently its out-of-place appearance among the more serious sports-racing machinery in the paddock prevented many from taking it seriously. Its C-type engine was said to be standard except for two 2in SU carburettors, but it kept ahead of a 300 SL Mercedes. Since an XK 140 fixed-head coupé is at least 220lb (100kg) heavier than an XK 120 roadster, assuming a similar weight reduction for both when stripped for racing, one would have again expected terminal trouble with its drum brakes. However, thanks to wire wheels and large holes cut in the front of the car to admit more air (they grew even larger after practice), it seems brake temperatures were kept at tolerable levels and the brakes held out.

Indeed, the car performed extremely well, lapping fast enough to find itself in twelfth place overall. Alas, it seems that there was a certain absence of forward planning in its pit as regards race strategy. David Murray of Ecurie Ecosse sportingly lent some assistance, although he may have had other things on his mind as one of his cars was in the process of winning the race. Unfortunately, after some sixteen hours of racing the XK 140 was accidentally refuelled one lap too early and five hours after this mistake the car was disqualified.

Assuming it had maintained its steady progress to the finish, it has been calculated that it would have reached as high as eighth place – three places higher than the XK 120 managed in 1951. Considering the XK 140's extra weight and the fact the XK design was now five years older, this was so nearly a truly brilliant achievement. The disqualification was a bitter blow, but in motor racing the rules have to stand.

Thus did the XK take part in three Grand Prix d'Endurance races at the Sarthe circuit. In 1950 three semi-works cars were entered with first-rate drivers, and in 1951 and 1956 a single privately entered car was driven each time by comparatively inexperienced drivers. The results achieved by the five cars were one breakdown, one eleventh place, one twelfth place, one disqualification while in twelfth position, and one fifteenth place. Allowing for the specialist nature of many of the cars they were competing against, these positions were really very impressive. And considering there was only one breakdown in a total of 9,500 miles (15,000km) of motor racing at an average speed of over 80mph (130km/h), the great basic strength of the Jaguar XK sports car was seldom better demonstrated.

BOB BERRY

To return to the early 1950s, one of the fastest XK 120s in the land would soon be Bob Berry's. Having raced his car, MWK 120, in a few events in 1953, he put on it the unused 1951 Le Mans lightweight body, LT-1, and reducing the weight even further, he campaigned the car the following year. He was assisted with race preparation by John Lea, another factory employee with considerable competition experience. This car had apparently shed about 650lb (300kg) when compared to a stock XK 120 and this had a remarkable effect on the brakes. Still using drums and disc wheels, fade was no longer a problem. Smells and smoke, perhaps, but no more fade.

Racing – 1951 and Beyond

DICK PROTHEROE AND ROBIN BECK

There seems to be a barrier of one ton that, if exceeded, condemns a drum-braked Jaguar to a career of brake trouble. The American cars fitted with LT-2 and LT-3 did have brake fade problems, but these were possibly heavier than Berry's car and since they were racing a couple of years before Berry they had neither the benefit of more up-to-date materials, nor direct access to the factory for (unofficial) help. In the early 1950s two other XK and driver

The extra weight of the XK 140 fixed-head coupé did not prevent some owners from enjoying themselves in competition, as this shot depicting a rather wide line at Prescott shows.

Nor indeed, would the bulk of the XK 150 deter them. This is an M. Fairsten at Goodwood in 1961.

113

Racing – 1951 and Beyond

combinations to be respected were Dick Protheroe in GPN 635 and Robin Beck in MXJ 954.

Protheroe was in Bomber Command in the war and had bought the car, an early alloy model, in Egypt. Upon returning to England, the car was given its now famous registration – presumably a deliberate choice? Very soon it would be nicknamed the 'Ancient Egyptian'. Protheroe was now a Leicestershire garage owner and it seems his ability behind the wheel was matched by his tuning skills. GPN 635 often displayed a delightfully battered appearance (from more than its fair share of prangs) but it would keep winning races into the 1960s, after Protheroe had flirted with an Austin Healey for the 1958 season. One of Protheroe's many successes was the over 2,000cc class in the three-hour night race at Snetterton in October 1959. Perhaps his stint with Bomber Command had given him some useful training for this race.

His next weapon was also an XK 120 roadster, CUT 6. This car was highly modified, using Protheroe's extensive experience with GPN 635, and in fact CUT 6 beat the newly announced E-type on more than one occasion. This must have greatly upset the proud owners of Jaguar's new state-of-the-art sports car, being unable to catch the far older model, but this was just what Protheroe wanted. The car's registration was, of course, one digit removed from Protheroe's next and more famous mount, the E-type CUT 7. Very sadly, this excellent racing driver was killed in the 1966 Tourist Trophy at Oulton Park, driving a Ferrari.

Although not quite as successful as Protheroe, Robin Beck also featured high in the results in the 1950s and into the 1960s with an XK 120 roadster, MXJ 954. This car was race-prepared by Dick Protheroe and would accumulate a series of modifications, eventually having its chassis shortened. A former rally car, it was unusually hacked about in appearance – but that did not stop it winning. By 1964 its 3.8 engine was producing a net 290bhp and this was another XK 120 that could severely embarrass the new E-types. One famous race that Beck did not win was the

Dick Protheroe in the 'Ancient Egyptian' in 1959. This car would end up more modified and considerably more battered-looking than this.

114

Racing – 1951 and Beyond

Protheroe leading Robin Sturgess at Club Corner, Silverstone, in October 1961. CUT 6 could keep ahead of the newly introduced E-types, even running in the very standard-looking guise as seen here.

Another XK 120 owner to prolong the racing career of the XK 120, Robin Beck at Silverstone in 1961. Judging by the position of the bumper support bolts, this car has had its fair share of fun.

115

Racing – 1951 and Beyond

1964 national Jaguar Drivers' Club meeting at Crystal Palace where he was eventually beaten home by Jackie Stewart in Eric Brown's D-type-powered drop-head XK 120, 1 ALL. Since Stewart was one of the very best drivers in the world at that time, perhaps Beck's car was really the faster of the two? Beck was not a man of traditional tastes, for apart from cutting and shutting the chassis of his XK 120 and altering the position of the headlamps, he would later campaign an E-type into which had been squeezed a seven litre Ford V8.

RHODDY HARVEY-BAILEY

Another very successful campaigner of the now quite elderly XK 120 was Rhoddy Harvey-Bailey in another early alloy roadster. The great thing about his mid-1960s success was that the car used was one of the original (unofficial) factory team cars, this one having been allocated to Clemente Biondetti, being 660043. Harvey-Bailey's father had bought JWK 650 in 1957 and with competition preparation by friend

A youthful Rhoddy Harvey-Bailey in 1965, furthering the competition career of the ex-Biondetti alloy-bodied car, chassis number 660043.

Dick Protheroe (was the cowl over the bonnet opening a Protheroe trademark?), Harvey-Bailey senior entered the Jaguar in some hill climbs. Once his son Rhoddy was old enough to own a licence, he began to drive JWK 650 competitively, very soon demonstrating exceptional skill and going on to record several outright victories and numerous class wins. When he began to compete in an E-type, again with help in race-preparation by Dick Protheroe, the engine was provided by JWK 650.

JWK 650 is now believed to be owned by Ralph Lauren, the fashion designer and a man who recognizes a thing of beauty. Today, Rhoddy Harvey-Bailey uses his very strong combination of considerable technical know-how and acute sensitivity behind the wheel (he always excelled in the wet) to design and market handling kits for various cars including, of course, Jaguar XKs.

JOHN N. PEARSON

By the time the 1970s arrived, it would have been almost foolish to predict that any XK could still be an overall winner, but a Birmingham driver helped to take the story of the XK another step closer towards its revered status.

Starting with a 1954 drop-head coupé that he raced with success in 1969 fitted with a 3.8 Mk X engine, he was to become known as John N. Pearson, to distinguish him from another John Pearson who was also a successful XK 120 racer competing at the same time. Pearson, an engineer, soon made a replica body shell in fibreglass for his drop-head, this now earning him the nickname John 'Plastic' Pearson. With the customary drastic modifications allowed at that time in the modified category, the car became quite unassailable in the early

Racing – 1951 and Beyond

1970s. Even fully modified E-types were left in its wake and just to rub salt into their wounds the XK, named Jessie, would soon wear the number plate 'I AM 21'. By the time a change in the regulations had ended the Modsports category in which this red beast raced, the driver was often simply known as Plastic Pearson. Did he ever appreciate the true scale of pleasure that he gave to XK 120 owners – as well as Jaguar owners in general – by winning so many races with nominally the oldest car on the grid?

It is only fair to state that at the time of all these victories, the car was no more an XK than it was an E-type. The suspension was now by Lola coil spring and damper units all round and rear axle location was substantially altered – as it would have to be without the leaf springs. Brakes were discs from an E-type and there were many other alterations such the radiator making its way into the boot to improve traction, the rear end of the chassis becoming a space frame aft of the axle, and the engine being shifted 2½in (60mm) further back in the chassis. All these modifications to Pearson's car meant its originality was almost non-existent, making one question whether it was really an XK at all. This did not matter one jot to the fans at the time of its dominance.

DAVID PREECE

The rearward shift of the engine in Pearson's projectile seems to contradict the claim that moving the XK 140's engine forward improved its handling. Since an XK 120 tends to understeer at the limit of adhesion, moving its engine forward would hardly make it handle better. In the late 1970s, there was yet another XK 120 that was wiping the floor with far more modern machinery. Incredibly, it was far less of a 'way out' machine than Plastic Pearson's frightening device.

A dentist by profession, David Preece moved from great success at rallycross to

David Preece, a dentist, extracting the maximum performace in the Oldham and Crowther car. At one stage in its development this car was apparently propelled by more than 350bhp.

117

Racing – 1951 and Beyond

win the Thoroughbred Sports Car Championship outright in the late 1970s. Between 1976 and 1978, he took 15 overall wins out of 24 starts. The car – prepared by Martin Crowther of Oldham and Crowther – was an XK 120 roadster, no surprise there. Registered WPU 207, it had to run with unaltered bodywork, so flared wheel arches and therefore huge tyres were out. Suspension also had to remain pretty well to the original design as well, so this car was far more of a real XK than one might imagine.

The car also competed in the modified category of the Thoroughbred Sports Car Championship – and yet was still far more original than the Pearson projectile, although the engine was certainly more powerful. Breathing through Webers with yawning 58mm throats, it produced a claimed 360bhp at 7,600rpm, at which the piston speed was around 5,300ft/min – an astonishing speed for a basic design that was then well over 30 years old. This XK 120 was faster around the tracks than even the Lister Jaguars, an extraordinary achievement.

THE XK 140 – A RARE SIGHT ON THE RACE TRACK

Apart from the impressive, but ultimately disappointing, performance in the 1956 Le Mans race, the XK 140 was seen very little in track racing, compared to the 120. No doubt the extra weight went against it, as well as that forward engine position, perhaps. This said, David Hobbs had some success in an XK 140 that was fitted with a semi-automatic gearbox as developed by his Australian father, Howard.

A serious attempt in 1959 at racing an XK 140, and a fixed-head coupé at that. Not many XK 140 and 150 owners went this far.

Racing – 1951 and Beyond

The start of a track race for a mixed bag of XKs, as a stage in a rally, Aintree 1958.

David Hobbs in his semi-automatic transmission XK 140 drop-head coupé, in 1960 at Oulton Park. Soon after, in this race, the car inverted and would later be re-bodied to resemble an E-type.

Racing – 1951 and Beyond

David Hobbs was a Daimler apprentice who became a Jaguar apprentice, by adoption, in 1960. In this year he achieved several wins in TAC 743, a drop-head coupé fitted with disc brakes. After altering the bodywork against a bank at Oulton Park, he moved on to race other cars, including Jaguars, for more than twenty years. Interestingly, TAC 743 was re-bodied at the beginning of the 1960s as a sort of imitation E-type, reminiscent of the way that an SS 100 was re-skinned by Uhlik to resemble an XK 120, not long after the later car's debut. (This recalls the fact that when the XK 140 first appeared, some XK 120 owners, keen to impress beyond their financial powers, had XK 140 grilles, lights and bumpers grafted on to their cars.)

In America, Bob Smiley raced a modified XK 140 roadster as late as the 1980s with

An as-new XK 140 drop-head, driven apparently by Eunice Griffin, awaiting the start of a hill climb at Prescott in 1956. The XK 120 just visible in front displays a novel position for the rear reflectors that had recently become mandatory.

Racing – 1951 and Beyond

The future Mrs Protheroe, Rosemary Massey, cornering hard in the 200 mile relay race at Silverstone in May 1959.

great success in Sports Car Club of America events, frequently beating cars that were half the Jaguar's age. He well deserved the American 'Jaguar Driver of the Year' award he received in 1984.

THE XK 150 – RACING FOR PLEASURE

Racing successes for the XK 150 are even harder to find. Not only was the car heavier, but its centre of gravity, thanks to that high scuttle and wing line, was a little higher than its predecessors. Certainly the brakes were enormously better suited to racing than were the previous models' drums, but then by the time the XK 150 appeared, the earlier and lighter cars were also being fitted by their owners with discs as well. Furthermore, the overall XK design was now entering its tenth year, so competing seriously on the track with a weight handicap was really out of the question. Yet another problem for the XK 150

Privateers who race their cherished cars are to be admired, for this is always a possibility. It happened at Mallory Park in the 1970s at Devil's Elbow, to Mr Fisher-Skinner and his XK 120 roadster.

was that by the time it was four years old, it had the E-type to contend with whereas the previous two XK models were only followed by heavier versions of themselves.

Nevertheless, some XK 150s did find their way on to starting grids, but none was raced with a realistic chance of achieving outright victory. Surely then, owners who competed with their XK 150s did so purely for enjoyment – and that is something often seen to be missing from motor racing today.

8 Rallying and Record Breaking

In the spring of 1949, the second left-hand drive XK 120 (the third of the three pre-production prototypes) was deemed fit enough to find out what speed could be attained in top gear. The car, fitted with very temporary-looking sidelights atop the front wings, was taken to Ostend in Belgium (via London where a carburettor fell apart), and was driven flat out on the Jabbeke highway, over the same stretch as the Gardner record car. Ron Sutton drove this bronze roadster – the same colour as the Earls Court Show car – while John Lea pressed the stop-watch. The outcome of this venture yielded results that must have caused some relief. Speeds attained ranged from 120mph (193km/h) to 133mph (214km/h), depending on the removal or addition of items such as hood, sidescreens, undershield, tonneau cover and windscreen. Mission accomplished, Lyons invited the press to an official demonstration run, at the end of the following May.

Chartering a 27-seater DC-3 aircraft, William Lyons accompanied a party of motoring writers to Belgium (was this the first such overseas jaunt?). With 670002 now repainted white 'to stand out' and now registered as HKV 500, Sutton managed to improve slightly on his previous runs. Averages of 126mph (203km/h) were recorded with 'full touring trim' – hood and sidescreens erected. With aero screens and tonneau cover in place, 133mph (214km/h) was attained. Upon returning from the last of the high-speed runs, Sutton pottered silently past the well-impressed pressmen at about 10 mph (16km/h) in top gear, thus demonstrating the docility and flexibility of the engine. In 1949, for a car that had just exceeded 130mph (210km/h), this was simply stunning.

THE XK 120 AS A RALLY CAR

As already explained, this very high top speed did not allay all the XK's critics, for they now cast doubts upon the car's ability in competition. No doubt with the aim of silencing them once and for all, Lyons set aside six quasi-works cars to compete on Jaguar's behalf, overtly to be entered by their drivers. As discussed in Chapter 6, five of these alloy-bodied cars would race on the tracks, but a sixth, chassis number 660044 and given the now celebrated registration NUB 120, was allocated to Ian Appleyard to compete in rallies.

Although the XK 120 showed considerable promise on the race tracks and won many, many races when modified in the hands of privateers, it only actually won a handful of top international races when in its (relatively) unmodified prime. When it came to rallying, however, it was at one time almost the essential requirement for the serious rally driver.

The Achilles heel of the 120, its brakes, proved less of a problem in a rally where flat out speed is not needed on every stage. It is also not possible to cane one's brakes as hard on a loose surface as on dry tarmac, although descending the mountains the

Rallying and Record Breaking

No problem with brakes in a sprint – this one at Bircotes, near Sheffield, in 1953. Presumably the driver is watching his fellow competitor.

fade problem would become a nuisance. At least in a rally, you get to stop every so often, when brakes can cool and also be adjusted.

The fact that the chassis was taken from the Mk V saloon meant that an XK's ground clearance was really very good for a sports car, and good ground clearance is always useful on the sort of poorly surfaced dirt roads that so often were, and still are, used for rallying. Furthermore, the great basic strength of the XK's engine, transmission and chassis frame would also be a powerful weapon in the gruelling conditions of a rally where the weight handicap that these robust items brought with them would be less of a problem than in a circuit race.

Ian Appleyard was, as many top racing and rally drivers, an exceptionally gifted sportsman. He represented his county at tennis, as well as his country as an Olympic skier. The acute sense of balance required for this latter pursuit would always come in handy for controlling a car on the limit. (It is no coincidence that most Formula One drivers are excellent waterskiers.) Having just married Pat Lyons, the daughter of his boss, Appleyard found himself with NUB 120 at his disposal. He had already had much success with an SS 100 loaned by Lyons immediately after the war, culminating in an Alpine Cup and Best Individual Performance in the 1948 Alpine rally. The following year he gained first place in the over 1,500cc class and second place overall in the Tulip Rally. This shows that becoming son-in-law to William Lyons

123

Rallying and Record Breaking

The XK's ground clearance was not that good. Rather a lot of wheelspin from the right rear suggests they may shortly be getting stuck. This XK 120 has twin exhausts.

was not the reason he acquired the XK: Lyons was far too hard-headed for that and it was Appleyard's talent that persuaded him. Once again, Lyons's 'investment' paid huge dividends for Jaguar, for Appleyard had enormous rallying success around Europe with the white XK 120.

His first rally sortie in the Jaguar open two-seater was in the 1950 Tulip where he forfeited a probable outright win by stopping 2in (50mm) short on a driving test at the end of the rally, resulting in a large penalty. This disappointment only caused him to try even harder in the future and after Best Performance by a Production Car in the Morecambe Rally, Appleyard prepared NUB for his fourth consecutive Alpine.

THE 1950 ALPINE RALLY

Appleyard had wanted to enter an XK in the 1949 Alpine Rally, but the car had yet to be ratified as a production model and so was ineligible. He therefore drove a Healey Silverstone and might have won a second Coupe des Alpes but for an agonizing wait at a level crossing. Thus the first Alpine an XK would take part in would be in 1950. Ian Appleyard and his wife Pat were accompanied by another factory-supported XK in the rally, for the unregistered Haines car also took part. An M. Barsley and the Swiss driver Ruef Habisreutinger also entered two more XK 120s, the latter ending this Alpine rally with his XK dangling over the edge of a gorge, while the former came home twenty-first in the general classification. A faulty master cylinder would give Nick Haines much unwanted excitement, too, before causing him and 660041 to finish the rally against a truck. Pat and Ian Appleyard fared better than this trio, although they certainly had their troubles as well.

The Alpine Rally was a particularly tough one; the fact that its pre-war title had been the Alpine Trial may be some indication of this. However, despite considerable physical discomfort in this, her first rally, Pat Appleyard persevered – she was William Lyons's daughter after all – and would go on to be her husband's navigator until he retired from international competition in the mid-1950s. Since the Appleyards had married less than two months before the event, some might say this baptism of fire was not the kindest thing to inflict upon one's new bride.

With his great experience of the Alpine Rally, Appleyard would carry out the detailed maintenance of NUB 120, which was duly fitted with a number of extras. These included two dashboard-mounted stop clocks and an extra horn, all to be operated by the passenger. The only modifications to the car apart from these bolt-on items were the larger fuel tank – it grew by 8½ gallons (40 litres) – and a manual override to that infernal Jaguar electric choke. (This modification must be the most common one carried out to regularly used XKs.) Finally, a number of spares would be loaded into the car – such as a spare coil, bulbs, radiator hoses, inner tubes and two spare wheels. Thus equipped, Ian and Pat Appleyard and NUB 120 set off for the start at Marseille.

On the way, they tried the car on some mountain passes and found to their consternation that the combined effects of the altitude, the very high ambient temperatures and the rather volatile French petrol caused vapour locks in the carburettors. It was decided to cut four holes in the bonnet to reduce the temperatures under it, as well as to run on the poorer grade fuel. This latter precaution would reduce engine power somewhat, but Appleyard decided that was better than being unable to restart a hot engine, as might be the case if the top-grade French fuel were used.

The Appleyards started off well enough, although they were somewhat perturbed to pass several wrecked cars in the first few hours. Just 50 miles (80km) from the end of the first stage to Monte Carlo, the XK threatened to run out of fuel. The fuel consumption had proved far worse than expected, thanks to all the slow hairpin bends, no doubt. By good fortune, a garage was found 4,900ft (1,500m) up the Alps and great distress and disappointment was avoided.

In the flying kilometre trial, it was encouraging that the four Jaguars took the first four places, but perhaps a little disappointing, considering the XKs' type-number, that the fastest of these Jaguar two-seaters, NUB 120, only reached 110mph (175km/h). On its 3.64:1 axle ratio, this meant that the engine could only pull 4,700rpm in top. This performance suggests that NUB's engine was pretty standard and can be partly blamed on the rather poor quality petrol in use after the vapour locking experience.

After a frantic search for two tyres to replace the rears that had worn down sooner than expected, they managed to track some down in the very short time available. This was at a similar level on the luck scale as finding the petrol station in the Alps. However, NUB 120 was soon struck by a gearbox that was reluctant to leave first gear.

This lock-up sometimes afflicted first gear on the old constant mesh Moss gearbox when great strain was put upon it. In a track race, a jammed first gear was unlikely to happen since bottom gear was seldom used after the start. However, Appleyard had been holding on to first gear downhill in order to give the brakes a rest. After about a mile and a half like this at about 30mph (50km/h), the car suddenly juddered to a standstill, fixed fast in first. After waiting for a few minutes, they gingerly set off and managed with a great effort to heave the lever into another gear. This, once again, was fortunate indeed.

Unfortunately, now avoiding bottom gear, Appleyard was taking very sharp hairpins approaching the top of the scenic but daunting Stelvio Pass. In second gear, the engine was turning too slowly and, thanks to the altitude, it finally stalled. With several more of these corners to go, it was either first gear or nothing. To the driver's delight, the gearbox seemed to working normally and they successfully completed the next two time controls without penalty. Once again, good fortune was quickly followed by bad, for the gearbox jammed in first gear again. Just before stopping for the night it freed once again, but Appleyard's confidence in lasting out the remaining two days and 620 miles (1,000km) of the event must by this time have distinctly threadbare.

A jammed bottom gear was possible because the first gear wheel, being non-synchromesh, is moved laterally along the shaft to engage and disengage and sometimes it would seize on its splined sleeve. The same thing can happen to the reverse idler that would sometimes seize onto its shaft. The second and third gear wheels, being synchromesh, are not moved back and forth for selection, so lock-up from this cause is avoided. A gearbox has no direct top gear wheel.

Nevertheless, the Appleyards were still not penalized when well into the penultimate day of the rally and therefore on course for an Alpine Cup. However, the average speed they had to attain over the Furka Pass to Gletsch was made very much harder by the number of cars and tourist coaches still on the road. They just squeaked into the time control, after Pat had worked overtime on the extra klaxon slung under the grille at the front of the car. (Imagine running a timed section of a modern rally in and out of holiday traffic!)

Setting off on the last timed run of the day to Chamonix with ever-present worries about that first gear, they took one of the very tight hairpins a little too rapidly in second gear. The consequent collision with a wall caused a very buckled front wing, which had to be pulled off the tyre with the aid of some burly spectators, who no doubt would remember this incident for many a year.

Pressing on with the remaining stages of the rally, the Appleyards still found themselves without penalties and therefore on target for a Coupe des Alpes. The going was very rough in places, especially as the average speed they had to achieve was very high given the condition of the road surface in places, not to mention the long diversion around a fallen bridge, this extra half mile or so (1km) not being taken into account by the Swiss organizers. With the XK fairly scrabbling around the corners, they reached all controls in time, one of them with just one second to spare. After setting the best time in both runs of the driving tests, Ian and Pat Appleyard found they had won the rally outright by making Best Individual Performance in the general classification. They also were awarded a Coupe des Alpes for a penalty-free run: surprising, perhaps, considering the battered front of the XK.

This was a very fine performance indeed,

Rallying and Record Breaking

The going was too rough for this XK, although it was more likely lack of traction than mechanical failure that saw the rope brought into play. Simms Hill, Exeter Trial, 1953.

but all the more impressive for the 'never say die' attitude of Ian and Pat Appleyard. Many times they must have felt deep down that their goal was a forlorn one, but they just kept on trying. It was a thoroughly deserved victory for car number 139, but more was yet to come from the duo in their white Jaguar roadster. Not only did the XK 120 win on race tracks, but now it had proved a winner in a top international rally. The demand curve for the XK 120 was becoming steeper by the week.

OTHER WINS IN 1950

Appleyard followed up this July performance by gaining a first class award in the Lakeland Rally and then winning the East Anglian Motor Club rally, both in September. For this win, he beat Gillie Tyrer in the BMW 328 streamliner that many see as the inspiration for the shape of the XK 120. In November, he and NUB 120 were placed second in the Motor Cycling Club version of the RAC Rally. This 1,000 mile competition saw Appleyard's now famous XK 120 just beaten by an MG TD – a chastening result for an XK fan to consider.

Meanwhile, on 24 October 1950, it was decided once more to show the public how fast the 120 was – as if they needed reminding. (Since 92 per cent of XK Jaguars made until the end of 1953 were exported, some of the patient potential purchasers in Britain probably did not want to be reminded.) The drivers to perform the feat were the man who won the XK's first race, Leslie Johnson, and Stirling Moss. The car was JWK 651 and they managed to

127

Rallying and Record Breaking

JWK 651 after averaging 107mph for 24 hours. Moss, wearing goggles, has fellow co-driver and car's 'owner' Leslie Johnson, on his right. Johnson has his arm around David MacDonald, the Dunlop tyre technician – presumably in gratitude for a safe run.

circulate the Montlhéry track near Paris for 24 hours at an average speed of more than 107mph (172km/h). This was all the more remarkable as the track at Montlhéry was in a rather poor state of repair and had become quite bumpy – not quite the right conditions for maximum speed.

Bar none, 1951 was the most important year for Jaguar in competitions. With the introduction of the C-type for track racing and its prompt win at Le Mans, it was now left to the XK 120 to prove its worth once again on the rally circuit. This it did with a vengeance, but first of all, in March, Leslie Johnson had decided to have another go at a flat-out run in JWK 651.

This time the idea was to see how fast the XK could go for one hour; it was considerably more modified, giving over 180bhp and fitted with an undershield, aero screens and a metal tonneau over the passenger's cockpit. An average for the hour of 131mph (211km/h) was managed at Montlhéry, from a standing start. This was about the same speed that HKV 500 had held for about twenty-seven seconds two years previously along the Jabbeke highway. It seems as if Johnson had acquired something of a predilection for setting high-speed records.

The following month, in the international Rallye du Soleil, Jaguar XK 120s came home first, second, third, fourth and sixth – a result to be emulated by the D-type at the 1957 Le Mans. The winning XK was driven by Henri Peignaux, Jaguar's main agent in Lyon. This was a marvellous result for Jaguar, but this rally did not have the cachet of the next one, the Tulip, which was held shortly after.

Appleyard won the 1951 Tulip Rally in NUB 120 and was followed home by Ruef Habisreutinger, the Swiss driver, who was partnered on this occasion by Horning in another XK 120. Appleyard followed this with first place in the Morecambe Rally, an event that saw many entrants in XK 120s. In July, he and NUB made the Best Individual Performance (there was no official General Classification) in the RAC

Rally, beating an amazing thirty-six XKs home, as well as some fourteen other Jaguars entered. Morgans were the second and third least penalized cars home, ahead of this whole string of XK 120s that included John Lyons – son of William Lyons and brother-in-law of Ian Appleyard – in seventh place. It would have been distinctly embarrassing had Appleyard not finished, given the fifty Jaguar entries in the event, since the remaining forty-nine of them would have effectively finished behind the two Morgans, being more penalized than the two cars from Malvern.

It seems that if you did not have an XK at your disposal for a 1951 rally, you had apparently already reduced your chance of winning. Indeed, the Yorkshire, Scottish, and Edinburgh rallies were all won by XK 120 entrants. Appleyard, it seems, was saving himself for his particular favourite, the Alpine.

Ten Jaguars were entered for the event and apart from the Appleyards, the factory lent its support to two other pairings. Nicknamed 'Gatso', the Dutch rally driver Maurice Gatsonides was partnered by a *Daily Telegraph* motoring journalist, Bill McKenzie. The third pair were Habisreutinger and Horning. There were also a few other XK 120s entered in the event, apart from the three covert factory-supported ones. These included those of Norton, Samworth, Soler and Sutcliffe.

The Appleyard car soon proved the fastest of the Jaguars present. Indeed, it was the swiftest car in the rally although it would be given a very tough fight by the Cadillac-engined Allard J2 of Godfrey 'Goff' Imhof. NUB 120 consumed not only its own, but two further sets of brake drums brought from Coventry in the (unofficial) support car, JWK 675, but also the front shoes and one drum from one of the privately entered XKs that had rendered itself out of action by striking a bridge. This makes a total of thirteen brake drums, which may represent a record consumption for one event.

The Gatsonides 120 suffered water pump failure and the subsequent overheating warped the head sufficiently for the gasket to be unable to seal the block to head interface. It seems that Gatsonides tried gallantly to make the last few controls in order limp home to qualify as a finisher, so that Jaguar would win the team prize. Alas, he did not succeed. (He would, however, go on to invent a major but highly unpopular contribution to road safety: a camera triggered by a speeding vehicle.)

The Appleyards and Habisreutinger reached Cannes without loss of marks, gaining them both an Alpine Cup, this being the Appleyards' third. Despite there being no general classification, Appleyard won the final timed tests and so the over 3 litre class, giving him Best Individual Performance. Quite why such a performance did not simply gain an entrant in the Alpine the title of 'The Winner', must only be known to the organizers. With the XK 120 of Soler finishing well up the field, Jaguar was assured of the Team Prize.

THE 1951 'MARATHON DE LA ROUTE'

In August 1951 Johnny Claes entered his factory-loaned XK 120, the more than well-used HKV 500, in the gruelling Liège–Rome–Liège rally, also referred to as the 'Marathon de la Route'. This event was notorious as a car-breaker and would only be overtaken as such by the East African Safari, in later years.

Claes was partnered on this rally by Jacques Ickx, the Belgian motoring journalist and father of the Formula One

Rallying and Record Breaking

driver and six-times Le Mans winner, Jacky. Claes and Ickx senior won the 1951 Liège–Rome–Liège without loss of marks, a remarkable achievement considering the severity of the rally – there were 69 retirements from the 120 or so starters. It was also remarkable as nobody had ever achieved a clean sheet before in the rally. (Perhaps Ickx's ability in gruelling rallies would rub off on his illustrious son, for Jacky would win the Paris–Dakar rally in 1983.) Second place went to Herzet and Baudoin, also XK 120-mounted.

After finishing fifth in the previous year's Liège–Rome–Liège in his SS 100, Jacques Herzet had a wonderful Alfa-Alvis style body made for it by Bidée, the Brussels-based coachbuilder. Herzet must have enjoyed changing the shape of Lyons's products, because his second place in the 1951 event was achieved at the wheel of a re-bodied XK 120, the work having been carried out by Oblin, a coachbuilder more normally associated with Ferrari. Attractive as it was, it was no match for the original. A third 120 made it home in sixth position, entered by Laroche and Radix. These results are particularly good illustrations of the basic strength of an XK Jaguar, and Claes and Ickx, with their first place, achieved one of the XK's most important wins of all time.

Ian Appleyard seemed to harbour an aversion to this particular rally but he was soon winning again. First success was in the London Rally in September where he managed to present the only penalty-free card. His wife Pat was occupied this time, not as his navigator – this was the journalist Gordon Wilkins – but as an entrant herself in the rally. Soon after this event, she did much of the driving alongside her husband in the Lakeland Rally, where she was accorded the Ladies' Prize.

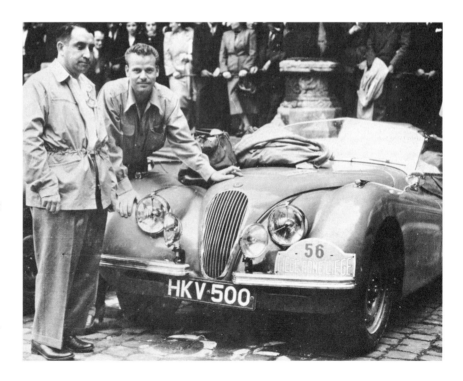

One of the XK's most important wins of all time, the highly versatile HKV 500 after winning the Liège–Rome–Liège Rally in 1951, driven by Ickx (left) and Claes. The spotlamps have been turned to illuminate the hairpins.

Rallying and Record Breaking

THE 1951 'TOUR DE FRANCE AUTOMOBILE'

The 'Tour de France Automobile' was a mixture of rallying and road racing. In September 1951 this hybrid event saw XK 120s finish first, second and third in the over 3,000cc class, although this translated to fifth, eighth and ninth in the general classification, with another 120 in twelfth position. Some consolation would be that three of the cars ahead of the fifth-placed Crespin and Hache were Ferraris, a very much more expensive marque altogether.

In 1951 the Appleyards had won the Tulip, the Alpine and the RAC rallies. Less importantly, from a prestige point of view, Appleyard had also triumphed in the Morecambe and London rallies. The outstanding rally team of 1951, Ian and Pat Appleyard were made guests of honour at a civic reception, hosted by the mayor of Leeds, their home city. If Appleyard was the year's outstanding rally driver, then the XK 120 was undeniably the outstanding rally car and in November it would head the results for 'Open Cars of over 3,000cc' in the Motor Cycling Club National Rally, driven by Grounds and Hay. Also in November 1951 an XK 120 managed a penalty-free run in winning the Welsh Rally, driven by C. Heath.

RALLIES IN 1952

We now saw the debut of the huge Mk VII as a rally competitor and Ian Appleyard

The Eight Clubs Eastbourne Rally in October 1952. Both drivers are displaying a certain amount of impatience, but what are the officials measuring?

Rallying and Record Breaking

would drive one, without success, in the Monte Carlo Rally of that year. There were a total of eleven Jaguars in this race. Returning to his more accustomed form of rally transport, NUB 120, he entered the RAC Rally in April. Surprisingly, perhaps, he was beaten into third place by another XK 120, driven by Jack Broadhead. The winner was Godfrey Imhof in a hideous J2 Allard.

In April Denham-Cooke won the Highland Three Days Rally in his XK 120 and in June he would win the over 2 litre open car class in the Scottish Rally. At April's Sun Rally at Cannes – the scene of an XK 120 benefit the previous year – Mr and Mrs Taylor won the over 2,500cc sports car class and were followed home by Habisreutinger and Horning, both pairings in XK 120s. At the Tulip Rally in the same month, with Ian Appleyard once again Mk VII-mounted, Habisreutinger and Horning moved up a place in their XK and won the over 3 litre sports car class. In May Pat Appleyard won the Ladies' Prize in the Morecambe Rally.

It was in June that the Appleyards entered the Alpine again. A penalty-free run would give them a Coupe d'Or des Alpes, for three consecutive clean sheets. Appleyard decided to try for this, rather than outright victory. As nobody had ever achieved such a feat before, this was a tall order indeed.

With NUB 120 now eligible to run in Special Equipment guise, it now had about twenty more horses to push it forward. Useful though this may have been, its brakes would have even more work to do, a job they were already finding too much for

A Ms L.D. Snow and friend enjoying themselves in the 1952 Eastbourne Rally while displaying an interesting line in mascots, spats and driving positions.

Rallying and Record Breaking

them. However, they were now self-adjusting at the front and with the permitted wire wheels, the extra horsepower was not to prove the problem it might have been. Maurice Gatsonides also drove a factory-prepared, if not factory-entered, XK 120. This was MDU 524, a car that would become the fastest XK of all time.

Amazingly, the 1952 Alpine Rally was won by a pre-war BMW 328, driven by the splendid couple Baron and Baroness von Falkenhausen, with Gatsonides second. The Appleyards were fourth, behind a Renault, but this was of little importance for they had achieved their clean sheet and thus their Coupe d'Or. The prestige gained for Jaguar by this particular achievement was considerable.

In the car-breaking Liège–Rome–Liège rally, Laroche and Radix brought their XK 120 open two-seater into second place. This car was yet another special-bodied XK, but their second place was a further example of the XK's inherent toughness, there being a 77 per cent failure rate in this year's event. Having left the road, the previous year's winner Johnny Claes failed to finish this time.

Also in August 1952, Leslie Johnson once again returned to Montlhéry with an

Mr and Mrs Rallying, after winning the first-ever Alpine Gold Cup for three consecutive penalty-free runs, all achieved in NUB 120, seen here at Henlys in Piccadilly, London W1. It is a legal requirement to cover competition numbers if driving the car on the road.

Rallying and Record Breaking

even more ambitious plan, this time using the second right-hand drive fixed-head coupé made, LWK 707. The car was less highly tuned than the previous Montlhéry record-breakers. It was fitted with a 24-gallon petrol tank, carried extra lighting and a two-way radio. This time he and three other drivers – Stirling Moss, Jack Fairman and Bert Hadley – circulated for a week, covering nearly 17,000 miles (over 27,000km) at an average speed of just over 100mph (160km/h). The weekly total would, however, be outside the rules of the FIA (the sport's international controlling body) since a cracked rear spring less than half way through the marathon did not find that spare carried in the car during the run. It was changed, nevertheless, while the clock ticked on. As a result, they captured several Class C world records up until the spring was replaced, but the record they had come for eluded them – officially.

The failure to capture the main world's record for the seven days did not really matter a jot: the Jaguar PR department made sure that the public would know what its XK had done when Shell prepared an official report on the excellent condition of the engine. The distance covered in one

'This Jaguar is the only car in the world having lapped 7 days and 7 nights consecutively at more than 161km/hour average, in covering 27,200km.' LWK 707 at the Paris Salon in 1952.

week was described as two years' normal motoring, but it is more striking to convert the distance travelled into a yearly figure: 876,000 miles (1,400,000km). Fanciful stuff perhaps, but this latest demonstration did much to illustrate the durability of the car and engine.

Boredom for the drivers on this monotonous circuit was always a problem, for which various unusual solutions were employed. On one occasion, Moss recalled finding Fairman and Johnson sitting in the middle of the track, playing cards. On another, he noticed that the gap through which the car ran, between the lap markerboard and the timekeeper's hut, was gradually diminishing lap by lap. Happily, the gap resumed its rightful dimension before anything too eventful occurred.

At the *Daily Express* Motor Cycling Club Rally in November, the Taylor husband and wife team won the open car class for second time in 1952 in their 120 roadster, although they were actually fifth overall. Stirling Moss could only manage thirteenth, although he won his class, for closed cars of over 3 litres, driving the two-tone fixed-head coupé LVC 345.

RETURN TO JABBEKE

On 1 April 1953 Jaguar decided to have another run down the Jabbeke highway. The XK 120 roadster, chassis 660986, registration MDU 524, was taken to Belgium. Driven by Norman Dewis, the dark green XK recorded an expected 142mph (228km/h) over the flying kilometre. This car was stated to be in Special Equipment

Stirling Moss in the Daily Express *Motor Cycling Club Rally in November 1952. His passenger is not preparing to jump out, but is possibly intent on seeing a braking point in one of the driving tests, although the driver appears rather unconcerned.*

Rallying and Record Breaking

Jabbeke 1953 and faster still. Driver Dewis is flanked by Joe Sutton on his right and Frank Rainbow. This picture reveals who worked on the car and who did not.

form and though sporting an aero screen, it looked remarkably standard. At this same test, a C-type reached 148mph (238km/h) – slightly slower than expected – and a Mk VII achieved 122mph (196km/h). This last figure is the most amazing, considering the silhouette dimensions of this fine-looking big saloon and that it exceeded calculations by a dozen miles per hour. 'Whatever next?' the public might well have asked itself of these speeds, but more – very much more in the case of the XK 120 – was not far away.

RALLIES IN 1953

The 1953 rally calendar had opened with the Monte Carlo Rally, for which the XK 120 was ineligible, even in closed form; cars with engines over than 1,500cc had to be four-seaters. Appleyard duly came second in a Mk VII, demonstrating his great skill as a rally driver. The next important rally would be the RAC, where an XK could be used. Before that, in March, the Lyon–Charbonnières Rally was won convincingly by Henri Peignaux, Jaguar's main agent in Lyon, in his XK 120 roadster.

Appleyard entered NUB 120 in the RAC, although he almost entered a C-type. NUB 120 had now travelled approaching 100,000 miles (160,000km), perhaps half of which had been at rally speeds: yet another example of the XK's ruggedness. Not only did Appleyard and his 'old' car win, but a clean sheet was recorded into the bargain. Fourth was Jack Broadhead, driving NMA 404, his XK 120 that Charlie Dodson drove so effectively in the 1951 Silverstone International Trophy meeting.

In May Appleyard would give NUB 120 its final outing in competition in the

Rallying and Record Breaking

NUB 120 in the process of taking Ian and Pat Appleyard to first place in the 1953 RAC Rally. This is just before the speed tests at Goodwood.

Morecambe Rally. Fittingly, the car won its last event, and won it well. It was also the last outright rally win for its driver. The rather tired cream XK roadster was then put out to pasture, into the works museum at the factory where it now lives, although it spent many years being admired at Beaulieu. Appleyard was not ready to retire just yet, for coming up next on the 1953 rally scene was the Alpine Rally, which had certainly become his speciality.

He acquired a new car, another white XK 120 roadster, chassis 661071. It even resembled NUB in its registration – RUB 120. Its rally career started off quite effectively, for the Appleyards won their class in the 1953 Alpine, although they were fifth in the general classification, beaten by three Porsches and a Ferrari, all smaller and more easily handled than the XK. They were still ahead of two more 120s, which came second and third in this over 2600cc class. More interesting was the fact that Appleyard won his sixth Coupe des Alpes for yet another penalty-free performance, although the difficulty factor seems to have been diluted this year, as Coupes were not only awarded to the two other XKs behind him, but to no fewer than twenty-three other entrants. These two XKs were an early fixed-head, 669024, driven by the Mansbridge husband and wife team and registered GFE 111, and another closed XK 120 driven by Charles Fraikin and Olivier Gendebien, the future Le Mans specialist.

137

Rallying and Record Breaking

Yet another married couple had success in an XK 120 in a rally, this time Air Vice-Marshal D. C. T. Bennett and his wife. Far down the general classification, they had the satisfaction of being top in the over 3,000cc class in the Evian–Mont Blanc Rally. Bennett would help get the little Fairthorpe sports car concern off the ground shortly after.

A better result was gained by Fraikin and Gendebien, who came second overall in their fixed-head 120 in the very strenuous Liège–Rome–Liège of 1953, behind 1951 winner Johnny Claes who drove the entire event single-handed, his partner unwell. Since Appleyard always avoided the rally because of the fatigue factor, this performance by Claes must rank as superhuman. It is very sad to note that he would only live a further two and a half years, succumbing to tuberculosis a few months after finishing third at the 1955 Le Mans race.

With the first European Rally Championship at stake, RUB 120 was given a green drop-head body that was modified to accommodate two vertically challenged persons in the rear. This was an attempt to enter the car in rallies as a four-seater where otherwise a committed Jaguar driver would have to use the Mk VII. Having got this 'four-seater' XK 120 through scrutineering in the Viking Rally in Norway, Appleyard was called to one side by the organizers. Although the car was eligible to start, it was not really in the spirit of the thing, was it, and would Mr Appleyard please enter with the Mk VII that he had brought along? Appleyard very sportingly did just that, a gesture inconceivable in today's modern world of motor sport.

Alas, more used to throwing his XK around on the loose stuff, he swiped a bridge and plunged into a river, his rally over. Some years later he described the Mk VII as quite unsuitable for the rally, being too large for the Norwegian roads. This does not explain very well how a Norwegian driver named Mathiesen in his first ever rally managed the fastest times on all the special stages – in a Mk VII.

An interesting postscript to this occurred when Appleyard made his track racing debut at Silverstone the following year, driving another Mk VII. He won the race, beating two more of Jaguar's largest cars driven by Le Mans winner Tony Rolt and Stirling Moss. Although the race's favourite, Moss, had his starter motor jam at the Le Mans-type start and was left well behind, it should be pointed out that all three drivers shared the same lap record in the race. This puts Appleyard's talent in perspective.

The final round counting towards the rally championship was the Lisbon Rally and was duly entered by Appleyard in his bespoke XK 120 drop-head coupé. He fairly hurled RUB 120 through the final tests, but was beaten into second place overall by a Porsche. With his main rival in the rally championship coming third, he was runner up to the title.

After Ian Appleyard announced his retirement from serious competition, RUB 120 was sold to Denis Scott, who competed with it in 1954. In the late 1960s, when XK 120 values were at their lowest point, it seems RUB 120 had reached hers too, and was scrapped. This would never happen now, of course, when XKs with no history at all are, happily, restored from almost complete wrecks.

SPEED RECORDS AND THE D-TYPE

A few weeks before the London Motor show in October 1953, a vastly more expensive car than the XK had just attained 151mph

Rallying and Record Breaking

(243km/h) along the Jabbeke stretch of Belgian highway – Jaguar's territory! A Spanish Pegaso, propelled by a supercharged V8 engine, had laid down a challenge to Jaguar in response to MDU 524's 142mph (228km/h) earlier in the year. Responding to this exciting provocation, a victorious return to the Jabbeke–Aalter run would help trumpet some welcome publicity for Jaguar on the eve of a show for which it had no new models to display.

With a Perspex bubble-type cover over the cockpit – Dewis was the driver again, perhaps a necessity because of his size – and metal covers over the rest of the cockpit, this car had very great attention paid to drag-reducing details. A 2.92:1 rear axle was fitted, the grille partly blanked off, and very special Dunlop tyres were fitted, which gave ultra-low rolling resistance. The engine details remain unclear to this day, but photographs of the car show breathers on the camshaft covers, rather like the D-type. Power must have been somewhere approaching 100bhp more than standard. Despite these intensive and speedily performed modifications, the results surprised even Jaguar.

With the car running on ordinary fuel obtained from the petrol pumps on the forecourt of a local garage – and the tank sealed by the official observer – the speed

Jabbeke 1953 again. Much concern (from even the young bystander) seems to be centred on the car while Lofty England reaches under the bonnet. Since more than 170mph was achieved, perhaps they need not have worried.

attained was almost exactly the same in both directions, for by a great stroke of luck there would be an utterly windless calm prevailing when the run took place. Jaguar were hoping for 155mph (250km/h), so were simultaneously amazed and delighted when just over 172mph (277km/h) was the figure reached. (Malcolm Sayer's slide rule must have contracted some form of virus.) Sadly, the car that performed this stunning feat was totally destroyed against a tree while on the way to a race meeting some years later.

Alongside the test of the XK, a sort of interim model between the C-type and D-type was tried out for top speed. Logged as the 'XK120C Mk II' – but now sometimes referred to as the 'C/D hybrid' – this car had many advanced features packed into its space-frame, including a triple-plate clutch, Weber carburettors and so forth. It also possessed far sleeker bodywork than a C-type. That it only managed to exceed MDU 524's speed by less than 6mph (10km/h) must make the XK's speed even more extraordinary.

Extraordinary is the right word, for the speed recorded by the XK 120 in 1953 was actually faster than J. G. Parry Thomas's World Land Speed record in 1926, in his fearsome and lethal contraption Babs, the 12-cylinder, 27 litre Higham Special.

Sometimes forgotten is the performance that day of a privately owned XK 120, which was allowed to run through the timing beams after Jaguar's own two-seater had finished its ballistic exercise. That its driver was allowed to do this can be explained by the fact that he was a Jaguar dealer from St Nicolas, west of Antwerp. He had bought his XK as an accident wreck and had rebuilt it with a few modifications. These were enough for even this privateer to beat the Pegaso, with a top speed of 154mph (248km/h).

It is interesting to notice that a shot of de Ridder's car travelling at speed shows it to be very close to the ground, far closer than a standard car and lower even than MDU 524. One benefit of this is that, because more of the tyres are covered by the bodywork, the car's total frontal area is reduced and thus the car will go faster. This leads to the interesting fact that a fully laden car, provided it remains level, will be faster than the same vehicle carrying only the driver. Another point of interest is that both the Jaguar and the De Ridder cars that day were themselves showing no discernible lift at the front. It is this 'rotation' that ruins a car's drag coefficient and that is clearly visible on photos taken of other XKs, at very high speeds. The first official 1949 Jabbeke maximum speed runs with HKV 500, for example, show it to be lifting considerably at the front.

The D-type was particularly prone to lifting its nose. At one Le Mans practice, J. Duncan Hamilton called at the pits to insist Lofty England do something about his D-type, because he said he could not see over the bonnet at maximum speed. Lofty fetched a cushion from behind the pit counter.

In 1982 a 'long nose' D-type had its coefficient of drag tested in the wind tunnel at the Motor Industry Research Association at Nuneaton. It was rather a surprise to find that it recorded 0.49 – apparently worse than an Austin Allegro. This disappointing figure obtained for the D was thanks mainly to the lift generated by the air stream. This is known as rotation, just as an aircraft rotates as it begins take off. It seems the D also could lift its tail as well, for how else could Tony Rolt have reported wheelspin at 170mph (275km/h) in the wet? Despite this positive lift, the D was always reported as feeling stable at maximum speed – obviously that distinctive tail fin worked.

Rallying and Record Breaking

The reason that the D-type cars were so fast down the Mulsanne straight was that they were so small. The D has a frontal area less than two-thirds of a Porsche 911 Turbo, so that its effective coefficient of drag (Cd x frontal area) is still only 80 per cent of the Porsche's, itself now regarded as the smallest supercar. A Formula One car has a Cd of about 0.7, but will exceed 200mph (320km/h) because of its great power and very small silhouette.

A postscript to this quite remarkable speed run by the XK 120 is that not long after this 172mph (277km/h) retaliation by Jaguar, the XK's last attempt at record breaking, Pegaso sadly ceased making cars. Perhaps they should not have announced their attempt as the 'highest officially timed speed ever reached by a sports car in Belgium'. It is always dangerous to pull the tiger by his tail.

RALLIES IN 1954 AND LATER

On the rallying scene in 1954 XK 120 successes were hard to come by, mainly thanks to the fact that nimbler cars were beginning to dominate. Nevertheless, J. Hally won the unlimited sports car class in the Scottish Rally in June and the following month the Alpine Rally saw an XK win its

A very new XK 120 drop-head coupé during the 1954 RAC Rally.

Rallying and Record Breaking

class, when Eric Haddon and Charles Vivian won the over 2,600cc category in their XK 120 roadster, RJH 400, chassis 661165. RUB 120 had been ahead of them in this class, but a broken rear spring caused by some very hard braking finished off the chances of new owner Denis Scott and his co-driver John Cunningham, who had himself won the over 2,600cc open car class in the Scottish Rally the year before in an XK 120.

Haddon and Vivian also won the over 2,600cc open sports car class in the Motor Cycling Club Round Britain Rally in November. Another XK, a 120 fixed-head coupé, won the same capacity closed sports car class, driven by Mr and Mrs L. Stoss. Two months before this rally, the XK 140 had been announced, but as a result of the appalling Le Mans crash the following June, many motor-sport events were cancelled, including the 1955 Alpine. Also aborted were a factory team of XK 140 roadsters that had been intended for rallying. We have to wait until 1956 for another significant XK rally success and this time it would be courtesy of an XK 140.

In March 1956 the Appleyards entered the RAC Rally in VUB 140, a white XK 140 fixed-head coupé with Special Equipment options. They managed to finish in second place in the general classification, two

A splendid shot capturing the period so well. An XK 140 drop-head coupé, with its row of 'medals', during the Little Rally in 1956.

Rallying and Record Breaking

Frank Grounds bending his two-tone XK 120 around a marker during the 1955 RAC Rally.

Ian Appleyard doing the same thing to his XK fixed-head coupé in the 1956 RAC Rally in which he came second to an Aston Martin DB2.

Rallying and Record Breaking

FDD 1 again, this time a real 'action' shot, during a 1956 rally, presumably in Scotland.

places ahead of the new 2.4 saloon, making its first competition appearance, driven by William Bleakley and Ian Hall. Ahead of the Appleyards was an Aston Martin DB2, driven by Lyndon Sims and Martin Ambrose. An XK 140 roadster had proved very fast in places: it was one of the cars built by the factory for the cancelled Alpine Rally. Driven by Ronnie Adams, it unhappily accrued so many penalties that it dropped out of sight in the final placings.

Although XK 140s entered by private owners did achieve some good class placings in rallies, none came close to VUB 140's performance in the 1956 RAC Rally. Perhaps the best result was to be found across the Atlantic in November 1956, when in the 'Great American Mountain Rally' a team of XK 140 roadsters managed a whitewash in their class.

The problem was that the XK 140 did not get the factory support and development for rallying that the XK 120 had enjoyed. Browns Lane were too busy with the D-types and, more significantly, with the compact saloons that would come to dominate the tracks as the XK 120 had done in its day. The same problems applied to the XK 150, but being heavier than its predecessor, it was to prove even more unsuitable for rallying, although some of its diminished dynamic performance, stemming from its increasingly archaic rear suspension, was restored with the advent of the S version.

It was an S that scored the only class win for an XK 150 in a major rally. Haddon and Vivian won the 3,000cc to 4,000cc class for GT cars in the 1960 Tulip Rally, in their XK 150S roadster, WLO 7, ahead of a Jensen. In the general classification they could only manage tenth place. This was the one and only significant rally result for the XK 150. Changes in the rules did not

144

Rallying and Record Breaking

The interior of an XK 140 prepared for rallying. Over twenty extra switches can be seen. Apart from the strange console next to the handbrake lever, also of note is that the car is automatic transmission, with the selector on the transmission tunnel.

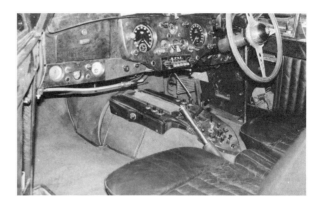

Also using their XK as it should be are these two naval types in a truly immaculate late-model 120 roadster.

favour large sports cars, for the E-type would have virtually no success at all in a rally. A rather different car would soon be dominant in rallying, one that could hardly be further removed from the XK Jaguar: the Mini Cooper. With the recent emergence of the historic, classic rallies, XK owners who like this sort of event can now enjoy themselves again.

In terms of rally victories and the consequent prestige brought to Jaguar by them, the XK 120 is certainly out on its own. Since it was the first of the type and since the XK 140 and XK 150 had almost no suspension development apart from rear telescopic shock absorbers, this is hardly surprising. Moreover, in terms of record-breaking, the XK 120 having managed over 170mph (270km/h) while looking relatively standard, there was little point in trying to improve upon such a speed with the XK 140 and 150.

145

9 Originality, Buying, Driving and Owning

The question of originality is an uncommonly emotive one. It is entirely up to the owner how original – or not – they want their car to be. As the last XK to be made drove gleaming off the production line in 1961, there is probably no such thing as a totally original XK left in the world. If there is, it will be a very, very low mileage car indeed.

No doubt many owners think they have an original car, but what about that battery? Or those tyres? Or that timing chain tensioner spring? The crankshaft oil seal? Or any of those items that were changed soon after production started for that particular model and the original specification part discontinued? Does the owner really feel better for having the original round disc pads on an XK 150, or the studless cam-covers on an early XK 120, which leak oil? What about running an early XK 120 without air cleaners? Original, yes; advisable, no.

One of the causes for this sometimes over-eager quest for original equipment is the phenomenon of the concours d'élégance. Started, perhaps, in the South of France in the 1930s for a bit of amusement after a rally or trial, the competitor and passenger usually displayed the car while sporting appropriately flamboyant costumes. This light-hearted piece of entertainment seems to have been taken up in America in the 1950s and become rather more serious and, frankly, over the top. Nowadays, the driver's and passenger's appearance are irrelevant and the competition for awards is very fierce. It used to be quite impossible even to drive these cars if you were to have a chance of winning, and a trailer would be used to fetch the car to and from the shows. As if to demonstrate that the cars were not meant to be used on roads, the entrant who actually dared to drive to a concours would usually polish inside the wheel arches on arrival. A car carried around on a trailer assumes the role of a painting and quite ceases to do what an XK still does rather well – take its owner for a spin.

While it is fully understandable that someone should want to restore their XK to first-class condition, it is when this quest prevents them from driving the car that one's comprehension begins to run dry. However, sense is finally prevailing and the XK Register of the Jaguar Drivers' Club will not admit any car to a concours that was not driven there. This is good news, although quite why anyone should want to demonstrate to others, by entering a contest, that they have the newest-looking car, is not easy to grasp.

Besides, an XK restored to concours standards, beautiful as it may be, has almost no character at all, being so clinically perfect in its make-up. It is so much more interesting to see a well cared for example that has obviously been owned and loved by the same person for some while, even if it may have one or two visible non-standard features to add to the owner's driving pleasure, such as an extra map light, spotlamp or luggage rack.

At the other end of the scale, is it right to

Originality, Buying, Driving and Owning

An over-gilded lily. The owner of this car must have shares in the company that makes fasteners.

put a Ford V8 engine with automatic transmission into an early XK 120? Or to flare the rear wheel arches enough to accommodate a Mk II axle fitted with extra wide low-profile tyres? How do you feel about three mascots on an XK 150 – on the bonnet and both front wings? All these examples have been seen, as well as one man's pictures in a club magazine of his DIY restyling of an XK 150 fixed-head coupé. You could quite believe that his car had enjoyed a violent rear end intrusion by a large motor boat.

Are these examples of how others wish to treat their XKs actually wrong? Where do you draw the line, at either striving for 100 per cent originality, or at modifying? Clearly, there is no set answer to this question, and it is a matter for each individual owner. Surely 'period mods' should be quite acceptable. That is, any mechanical alteration that saw Jaguar parts that were in production during the run of the XK series fitted to older models should not be frowned upon, such as the appearance of XK 150 disc brakes on either of the previous models. Whereas the fitment of, say, an AJ6 engine from the late 1980s, or even a non-Jaguar power unit, should merely be regarded as rather a shame. Perhaps a rough guide could be, as in so many things in life, not to go to extremes: either in overly modifying, or in an obsessive search for some apparently divine, immaculate perfection.

At the same time, the outward appearance of the XK is surely best left unaltered. After all, now that many facets of its performance are today matched by the most

Originality, Buying, Driving and Owning

It is always nice to admire the simple, uncluttered lines of the XK 120 roadster. Badges became popular after the war, when owners would award themselves 'medals'. At least this owner was brave enough to hill-climb his car. Prescott 1955.

dreadfully humble modern cars, an XK's looks are now its strongest asset. This said, suitable 16in tyres can be rather expensive: over £700 for four Pirelli Cinturatos at the time of writing was considerably more than the cost of tyres for a rim one inch smaller. Furthermore, fitting 15in wheels gives a much wider selection of tyres. Also, to run an early XK 120 without indicators, while not illegal in the UK, is inadvisable in modern traffic conditions.

It is interesting that it seemed to be the aim of almost everyone who bought an XK 120 or 140 with disc wheels to change them to wires, and the author was no exception. However, the XK 120 roadster actually looks better with disc wheels and its rear spats in place. Such a guise enhances the smooth, graceful lines of the car, while the fitment of wire wheels tends to break up the flow. This happens for three reasons.

First, the spinners, while delightfully aggressive, have a knobbly appearance that does not suit the car's stupendous grace. Second, without the wheel covers you can see the large gap between the tyre and the rear wing and this is not an attractive sight, especially on a car riding on its original-sized 6.00 x 16in covers. Third, you can see the stay holding the rear wing and this just looks wrong. This all goes to show that William Lyons's styling of the first XK was, quite simply, right first time.

It is beyond the scope of this type of book to list every production change to all the XK models, but a list of the significant

Originality, Buying, Driving and Owning

visual changes that occurred is given in the Appendix. You could argue that it must be obvious what colour a car was originally or whether it was made with chrome sidelights or not, but when a car is around forty years old, all manner of alterations and part-swapping may have long since been inflicted upon it. If an owner really is worried about the originality of the starting carburettor switch plunger, or whether the second gear synchronizing sleeve should or should not be fitted with a stop pin, either a parts manual or a psychiatrist should be consulted.

CHOOSING AN XK

Which XK should you choose? If you have never owned an XK before and, like jazz, you like what is there but do not know what to buy, perhaps the choice can be narrowed down with a little reasoning. It all depends on your needs, of course.

If the main priority is beauty, then most people would agree that the XK 120 in Super Sports guise is the car to seek, closely followed by the same car in fixed-head coupé clothing. The XK 120 roadster is also the model with the lion's share of XK history, and this certainly can be appealing.

If you buy an original XK 120, then the performance of the early steel roadsters is leisurely. The car tested by *The Autocar* in 1950 accelerated to 60mph (100km/h) more slowly than a modern 1.6 litre Nissan Sunny, which has a similar top speed to the XK as well. The later Special Equipment 120s were rather faster, getting into single figures for the dash to 60mph (100km/h) and managing a top speed true to the car's type number. Unfortunately, the XK 120 has very poor brakes if you want to use the performance, even that of the early cars, to the full. Cockpit space is cramped for tall people and there is little scope to rectify this problem in the roadster model.

The XK 140 may not have the simple, uncluttered beauty of the earlier car, but it does retain the XK 120's overall shape – at least in roadster and drop-head form. If you require space for small children, or are simply on the tall side, then the XK 140 in its two coupé forms is a far better bet than the 120. Although not immune to fade, its brakes are slightly less prone to the problem than the 120. Perhaps this is partly due to the slightly smaller friction lining to rubbed area ratio of the 140's brakes, compared to the first XK 120s?

When driving an XK 140 for the first time in the rain (and this applies to the XK 120 as well), it will immediately become apparent that the wipers are very nearly useless. Driving the roadster in wet weather could put you off XKs for life. The drop-head makes a far better convertible than the roadster and it is no wonder that the XK 150 roadster was a far more civilized machine than the previous two, or that the true roadster Jaguar disappeared altogether with the advent of the E-type.

With overdrive and a C-type cylinder head, the XK 140 is an effortless high speed cruiser. It was also the first XK to be able comfortably to exceed 120mph (195km/h), even if it could not approach the speed its name implies. Beware of stories of astronomical top speeds for this 210bhp (gross) XK, for a speedometer over-reading by 5mph (8km/h) at maximum speed is quite normal and what many owners seem to forget is that the radial-ply tyres, which are so often fitted to XKs today, do not grow with speed as did the old cross-plys. This will account for a further 7–8mph (11–13km/h) in speedometer optimism. Coupled with this is the fact that radials usually have shorter sidewalls than the old cross-plys. This reduction in the rolling

Originality, Buying, Driving and Owning

> **Calculation For Speedometer Check**
>
> You need to determine the differential ratio and the rolling radius, this being the distance from the centre of the rear wheel to the ground, with the car at normal laden weight, tyres at correct pressures and the car on a level the surface.
>
> M = mph/1,000revs in direct top gear
> R = Rolling radius (inches)
> D = Differential ratio
>
> 1,000 rpm engine speed yields
>
> $\frac{1,000}{D}$ rpm at the rear hubs, which is
>
> $\frac{(1,000 \times 2\pi R)}{D}$ in/min road speed, or
>
> $\frac{1,000 \times 2\pi R \times 60}{63360 D}$ mph, thus
>
> $M = \frac{R \times 5.95}{D}$ ie
>
> Mph/1,000 revs = $\frac{\text{Rolling radius} \times 5.95}{\text{Diff ratio}}$

radius and therefore circumference of the tyre is not very much, but can amount to around 4–5mph (6–8km/h) at 5,000rpm in top gear. Thus you can easily have a speedometer over-reading by 18mph (29km/h) at maximum speed.

To this you must attach the fact that claims as to how fast a particular car has gone are subject to more exaggeration than tales of fuel consumption achieved and the size of fishes that got away. This is why yarns about XKs that somehow displayed top speeds rather faster than the factory cars should be treated with disdain. For a check on XK speedometer accuracy, a simple method of calculating the true top gear speed per 1,000rpm is given above.

The XK 150 is the roomiest, smoothest-riding, fastest and best-braked XK. It is also the heaviest and it looks it. As with the XK 140 it has more leg room than the 120, as well as more width in the cockpit than either thanks to the slimmer doors. Because it is the heaviest XK it rides the best, because a bump will deflect a lighter car more easily than a heavy one. With its vastly superior brakes it therefore offers the most comfortable and relaxing way to travel. This is particularly true of the roadster version of the 150 with its wrap-around windscreen and wind-up windows. In 'S' guise it has a respectable performance even by today's standards, particularly from about 60mph (100km/h) onwards.

To look at the cars from a body style viewpoint, the roadster versions of all three XKs are the purest, least compromised and best-looking. However, the buffeting experienced by the 120 and 140 open two-seaters on a fast run with the hood and side-screens off is very unpleasant – so much for the joys of wind-in-the-hair motoring. In the winter, the side of one's face becomes paralysed, one ear will stop working and one's right hand (if a right-hand drive model) can simply freeze solid, being clamped rigidly to the wheel at the end of the journey. Such delights are avoided in the 150 roadster. One trick of the XK 120 and 140 roadster owner is to run hood-down, but with side-screens in place. It is surprising how warm the cockpit can be, assuming one's early XK 120 has a heater, of course. Despite these problems, the roadsters are usually the most expensive of the three styles to buy.

If you want open air motoring, but find the roadsters a bit Spartan, then the drop-head is an obvious choice. Indeed, the drop-head is really the best of both worlds, a car that has saloon-like comfort with the hood up but can be converted in to a fully open car in just a few seconds. This trick cannot

Originality, Buying, Driving and Owning

be performed by the roadsters. However, the drop-heads are generally a fair bit heavier than the roadsters and with the hood dropped, they always look ungainly. The XK 140 and 150 drop-heads also have the addition of two rear seats. Although very small, they can be used to transport similarly small passengers, or even an adult contortionist.

Can the term 'sports car' ever be applied to a vehicle with a metal roof? Many have questioned the point of making two-seater saloon cars, for that is essentially a fixed-head coupé. On the other hand, they can be the most affordable XK type, as well as the safest, of course. They will also have the highest true cruising speed, thanks to the reduction in wind noise and rain intrusion compared to the open two-seaters and, to a lesser extent, the drop-head coupés. They are also the most aerodynamic, so should have the highest top speeds, although this would in practice be hardly a useful increase over the open cars – perhaps two or three miles per hour.

The above may lead you to think that the 3.8S XK 150 roadster is considered the most desirable XK – and there are many who

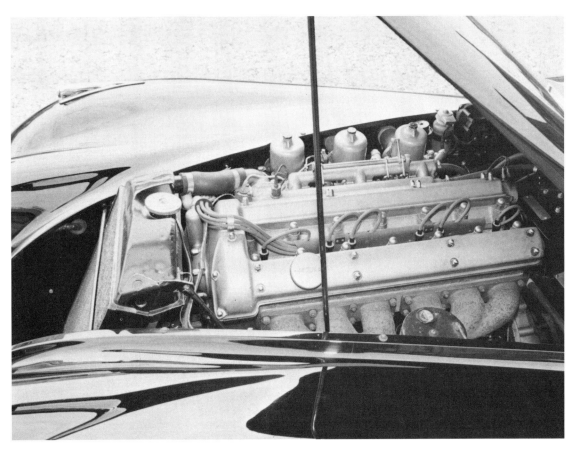

A triple-carburettor engine can be fitted into an XK 120, but the job is not for the faint-hearted as it involves much modifying.

Originality, Buying, Driving and Owning

would agree with this notion. If you are not repelled by modifying, then perhaps an XK 120 roadster with the mechanicals from a 3.8S XK 150 is the ultimate XK, as depicted on the front cover of this book. Such a package gives the best in looks, brakes, performance and light weight for good handling and fuel consumption – assuming, of course, that you can fit into it well enough to drive it properly. There is still the problem of fast driving in an XK 120 roadster in entirely open mode. The author did not have his for long before beginning to appreciate that those owners who wore flying helmets and suchlike were not posing.

Halogen conversions for those Lucas PF770 tripod lights on the XK 120 are now available, so that is another problem solved without altering the appearance. Although not really a 'period mod', the A-type compact overdrive will fit behind the all-synchromesh gearbox on an XK 120, so you can have the benefit of a better gearbox and five speeds. Its helical first gear is also quieter than on the older gearbox.

DRIVING AN XK

An XK 120 thus equipped can remain quite original in appearance and yet will see off many a supposedly fast modern car in a straight line. Taking on a current car on the twisty bits is a different game altogether, however. What, then, is an XK like to drive today?

The answer is that, if original, it is just the same to drive as it was when it was

This XK 120 has had two very useful modifications. The twin exhaust has been routed through the chassis, greatly helping ground clearance, and the compact overdrive unit has been fitted, giving five gears. Also visible are a much heftier anti-roll bar and the traditional oil leak from the back main bearing.

made. The trouble is, we have all driven much more modern cars and our perceptions have changed tremendously. The first thing to notice is the large size of the steering wheel, the second the driving position: your legs stuck straight out in front and your feet only 2–4in (50–100mm) lower than your posterior. Odd at first, this begins to feels quite normal as 'muscle memory' takes over.

Moving off, the steering on all models is surprisingly light when on the move, although it can load up quite a bit when cornering hard. The gearbox can be very sweet and, with a familiarity that is only truly fully developed after perhaps 20,000 miles (32,000km), it can be made to change up or down very rapidly. This explains why road testers, even driving new cars with unworn gearboxes, were unable to get the best from them in a few days' acquaintance. The Moss unit has a sweeter change than the later gearbox introduced in 1964. This gearbox has more inertia thanks to its larger internals as well as the baulk rings now fitted between the hubs and the constant mesh gearwheels. Jaguar's partial answer to this problem was to fit an extremely heavy gear knob to help 'throw' the change through more easily.

Smooth as the change can be when in good order, the Moss unit can wear out its synchromesh: many twenty year old XKs bought very cheaply in the late 1960s and early 1970s had very poor gear change quality. Because of the strength of the older gearbox, rebuilds to rectify a total failure were uncommon and because of the daunting task of removing the gearbox anyway, the fitment of new synchromesh parts was rare. As with any manual gearbox, second gear is the first to wear out its synchronization parts and a slurred grating sound would accompany any attempt to change from first gear too rapidly. Changing down into second often demanded a quick blip of the throttle in neutral before it would go in at all.

Merely renewing the springs under the ball bearings in the synchromesh hubs was quite an effective way to restore lost gear-change quality. New, stronger springs delay the sleeve in sliding off the hub and this in turn forces the cone clutch faces of the hub and gear-wheel to rub harder together, thus giving more chance of synchronizing the speeds of hub and gear-wheel before the sleeve pops off its ball bearings and into engagement with the dog teeth of the selected gear.

Many used to put up with the poor gear-change of the XK because if not in a tearing hurry it was possible to let the engine pull from absurdly low speeds in top gear without complaint. The tremendous torque given by the engine, even at low revolutions, is of course another salient feature of a drive in an XK, and it makes the car into a sort of one-gear automatic for any speed above 20mph (30km/h). It would still drive rather briskly using top gear only, once 30mph (50km/h) had been reached. This effortlessness, allied to the long-legged nature of the car's performance if an overdrive is fitted, continues to be one of the XK's enduring attractions, even in the 1990s. As for the car's roadholding and handling, such enormous strides have been made in these areas that any comparison with modern cars is rather unfair. Such comparisons are also unavoidable.

Roadholding on an XK is not remotely in the same league as a modern car and there would have been something very wrong in the progress of car design if it were. The main difficulty is revealed when you try to corner enthusiastically on a bumpy road. An XK does not like bumps in mid-corner, and even going at 30mph (50km/h) a degree of alertness is sometimes required lest the rear of the car should hop out too far on a bumpy bend. Some people rather like the idea of having to control their machinery like this and while a modern car generally discloses no bad behaviour in its roadholding until thoroughly dangerous speeds are attained, entertainment in an XK can start at much lower velocities. On the other hand, if you still want to corner fast in an XK, but lack this mildly masochistic element, the back axle will have to be modified by means of some kind of lateral location, such as a Panhard rod. Axle tramp can be dealt with similarly by fitting radius arms.

The front suspension, while truly first class in its day, is not quite up to modern standards, where all manner of bumps and dips are simply ignored. Some bump-steer can be experienced although anyone who has put rack and pinion steering onto an XK 120 may find this has become far worse. The problem lies in the rack sitting too high in relation to the front suspension. The rack will need to be set down inside the XK 120 chassis a little so that, viewed from the front of the car (in its 'normal riding position'), the steering arms and top wishbones are parallel.

Aside from behaviour on bumps, the grip of the car on a smooth surface will be found to be quite good and a determined XK driver can surprise the driver of a more modern car by keeping up. On the other hand, it would not be true to state that a determined XK driver could leave the modern car behind, were even a mildly interested driver at the newer car's wheel. Thrown hard at a corner, an XK will roll a great deal and if fitted with cross-plys, it will make some very loud protesting noises to tell you that it is about to slide. Happily, abrupt behaviour is not one of its vices, although an over-application of throttle in the wet in first or second gear might give a driver brought up on front-wheel drive a sudden increase in pulse rate. All is not lost if this happens, for the XK will come back into line simply by a lift on the accelerator and an armful of opposite lock – sometimes referred to affectionately as 'oppo', presumably by those rescued by its timely application.

This fairly benign behaviour of the XK when pushed to the limit is in marked contrast to the next Jaguar sports car, the E-type. This car displayed the early Porsche 911 trait of sudden lift-off oversteer, which rewarded those who over-reacted to a tail slide by snapping the other way with great suddenness. One long-term XK 140 roadster owner bought an E-type – very soon afterwards he demolished it completely by driving into the tunnel at Heathrow airport (he missed the hole, but was happily undamaged himself).

For an XK that is considerably improved in its handling, but without loss of comfort, there cannot be many companies better able to help than Harvey Bailey Engineering (see later in this chapter). The author has driven a modified XK 120 that had not been pleasing its owner in its handling. After a trip to Harvey Bailey Engineering, the owner described its behaviour as 'transformed'. Unable to generate enough speed, or courage, fully to verify his claim, the author can at least report that the ride comfort, even with Konis all round and stiffer than standard rear springs, was still better than his wife's company car, a pseudo-sporting 1995 model.

The overall comfort, ride included, amazed everyone in the late 1940s, and it is still a surprise today. One secret is to eliminate all rattles and squeaks, because a car that rides the bumps silently will disturb its occupants far less than one that crashes over them. This gives the feeling that the first car is smoother, even if the suspension is identical for both.

Originality, Buying, Driving and Owning

On a test drive, always try the brakes; if the car is the first of the two drum-braked models, the effect could be very impressive. From urban speeds, the brakes should work well enough, enabling the car to be pulled up very smartly. The handbrake should be able to lock the back wheels from this speed and this is a feat the author has never managed on any disc-braked car. Trying the brakes from very much higher speeds, especially if they are already hot, is a different story, as has been discussed, although the fade problem can be alleviated quite well with a little judicious modifying. Apart from the fade problem, drums have another drawback: snatch.

In damp conditions the first stop of the day can result in the most frightful grabbing. On some misty mornings the first application of the brakes, however light, causes the car to halt with an abruptness that must cause any witnesses to think the driver a complete novice. The second application of the brakes would be quite normal. The reason for this sometimes embarrassing phenomenon might be a combination of surface rust formed on the drums, swollen linings or swollen brake seals.

As for the more serious problem of overheating brakes, linings can be bought that are much more resistant to fade and the ensuing higher pedal pressures can be reduced by the fitment of servo assistance. Wire wheels that the majority of owners seem to fit nowadays will help the escape some of heat from the brakes, which is what causes the problem in the first place. However, if going to all this trouble, perhaps it is better to fit brakes from the XK 150 and be done with it.

The XK 150 was fitted with drums in its standard form, but almost no 150s were sold that were not either Special Equipment models, or, later, the S-type version. Driving the 150 at sedate speeds, you would not find the brakes any better than the previous model's drums. Indeed, the remote servo, being some distance from the master cylinder on a right-hand drive car, can give rise to a rather dead feel and no appreciable stopping action at the beginning of its stroke. This slight delay can be very disconcerting to someone used to the immediate action of a modern braking system. Once again, familiarity can make it seem perfectly normal.

When you start to travel faster the all-disc system of the XK 150 reveals itself to be light years ahead of the previous cars. At the sort of speeds that the S versions could attain and, more significantly, the frequency with which the high speeds could be re-attained after slowing, the discs are positively essential. Had Jaguar's discs not appeared when they did, an XK 150S with its performance and weight increase over the first 120, but still equipped with drum brakes, would be a frightening prospect.

REPAIRS AND MAINTENANCE

Your chosen XK can offer varying degrees of comfort, performance and braking ability. All versions will have similar cornering capabilities, although a lighter car will always be able to change direction faster than a heavier one. Since the XK's suspension was virtually unchanged throughout its production life, this suggests that the 120 will corner better than the other two. Behaviour on race tracks by the three types tends to confirm this.

What all three will also offer is a very smooth engine with a super-abundance of torque at any speed and effortless cruising at 70mph. Allied to this are all-leather seats and dashboards that are still better to look at than many a more modern Jaguar, alas. The poor ergonomics of which

Originality, Buying, Driving and Owning

These rear discs are about to be replaced. A new set will restore the 'bite' that has gradually been lost as the discs have deteriorated to this condition.

the car is accused will present no problems once the position of each switch is learned – muscle memory in operation again.

Fuel consumption can be summed up by stating that an XK will return two or three miles per gallon above or below the twenty mark (between 12 and 17 litres per 100 km). Extreme conditions, such as commuting in very heavy traffic or a very long motorway run in an overdrive model sticking resolutely to the speed limit, might see the consumption vary from single figures to the mid-twenties. Some XK owners claim that their cars regularly manage 30mpg (9.4 litres per 100km). Either their speedometers need re-calibrating or they do an awful lot of freewheeling.

By current standards, oil consumption is dreadful: two hundred miles per pint is quite normal. Oil consumption like this really is archaic nowadays, indeed it is not unusual for a modern car to use none at all between oil changes. The XK handbook advised the owner to change the oil and filter every 2,500 miles (4,000km). With the 29-pint (13-litre) early XK 120 sump, engine lubrication is not cheap, although modern oils will allow a change every five or six thousand miles (3,000 to 4,000km).

The oil pressure is always a good guide as to the overall condition of an XK unit. This should be 40psi at 3,000rpm with the engine hot. It is absolutely useless checking the oil pressure unless the engine is at full operating temperature because it can drop by as much as 50psi from cold to hot.

Originality, Buying, Driving and Owning

A subject of interest to the buyer of any car of any age is its reliability and the good news here is that the basic components of the XK are very strong. Its engine, gearbox and rear axle are very tough, as you might expect since the early cars won so many races and rallies. A properly maintained XK engine, with oil changes at the recommended intervals, will often last into five figure mileages without major overhaul. However, misuse and neglect can shorten the lives of any component and a Jaguar's legendary bargain status when around fifteen years old meant that many fell into the hands of owners who just could not afford to have them serviced the way the factory intended. This is true of all the more expensive cars. In consequence, all those lucky enough to buy a Jaguar for the price of a far smaller car, sometimes wondered whether the XK engine was as long-lasting as generally claimed.

The bottom timing chain is the engine's weak point. It often stretches to the point where failure of the tensioner occurs, followed by failure of the chain itself and an immediate and terminal intimacy of inlet valves and piston crowns. A worn bottom chain can often be heard tinkling inside the cover and timing covers can be almost cut through from the inside by a slack chain whose tensioner has long since dropped into the sump. Replacement of a worn chain means taking the head and sump off, but many garages used to cheat by jacking the head up very slightly and removing the cover using one or two other dodges. This may save time in the short term, but head gasket failure seems a certain consequence.

Valve guides are often worn and this is one of the main causes of high oil consumption, along with worn piston rings and leaks – the rear main bearing oil seal is a persistent offender on some cars. Worn rings are easily identified, since blue smoke will be emitted when the engine is hot and accelerating hard. If the guides are worn the blue smoke appears on the overrun, sucked down the guides into the combustion chamber when the throttle butterflies are closed.

A good way to test whether the guides are excessively worn is to reach a high engine speed in second gear and then lift off the throttle completely, letting the engine speed die down slowly with the road speed. All the while staying in the same gear, you then re-apply the throttle and if the guides are worn, a cloud of blue smoke will be emitted from the exhaust. In this artificially severe overrun condition, the engine is being driven by the rear wheels and so is rotating faster than it would be under a normal closed-throttle situation. The pistons are still trying to pull in air (and petrol), but unable to do so with the throttles shut, they create a partial vacuum in the cylinder that draws in oil down the guides.

Crankshaft failure is rare enough to be called non-existent, although one Jaguar XK owner, who drove his car from London to Brighton with his sump full of flushing oil, seemed shocked when his main bearings and big ends were ruined. A properly maintained engine will very seldom suffer head gasket failure (owners of early 4.2 litre engines will wince at this point) or run its bearings.

High mileage power units will often overheat, however, either as a result of a build-up of silt in the block, or a furred-up radiator core. It can be alarming when, stuck resolutely in a traffic jam, you watch the temperature gauge needle creeping into the oil pressure gauge.

Because the XK engine is generally so smooth, powerful and lacking in any flat-spots, it may have gone quite a long way off tune when in the hands of an inexperienced

Originality, Buying, Driving and Owning

With its EN16 steel crankshaft and massive seven main bearings, the great strength of the bottom end of the XK engine can be seen in this photograph. The connecting rods behind are unfashionably long today, where shorter strokes and wider bores are the order.

owner. The remarkable thing about the XK engine is the fact that the only special tool essential for maintenance is the camshaft sprocket timing gauge. This is a long name for a very simple and inexpensive part. This is in sharp contrast to a modern car that is bedecked with infernal splined keys and strange bolts. It is not in the range of this book to provide information normally given in workshop manuals, but setting the ignition timing is not difficult. Strobe light timing guns are quite unnecessary, since the fine adjustment is always arrived at by trial and error – just short of pinking is the usual setting. Those cars with 9:1 pistons may have trouble with today's petrol and running on unleaded or super unleaded is ill-advised, whatever the compression ratio of your XK.

Setting the mixture on the carburettors does not require elaborate vacuum devices and other fancy equipment, although a trip to an MOT centre might be advisable to check the exhaust emissions after making adjustments. Making sure that all carburettors open at exactly the same time is something that seems to be usually neglected. If they are synchronized in this respect as well as making the same hissing sound at tick-over, the improvement in the smoothness of the engine's pick up can be very satisfying. Compared to a modern car,

Originality, Buying, Driving and Owning

A later-type XK 120 engine that will soon be receiving a complete overhaul. Note the blanking plate for the obsolete oil sender, located in the bottom centre of the sump.

the spark-making and fuel systems are simplicity itself.

As for the much maligned Moss gearbox, teeth broken off first gear are not unheard of. Being straight cut, the teeth on bottom gear are not as strong as those on the all-synchromesh gearboxes that, being helical, are longer. Furthermore, helical gears have a 'wipe' factor in their engagement and disengagement as they drive each other, and this makes for quieter operation, generally free of the slight 'chatter' that can occur on straight cut gears. A gearbox run low on oil will naturally become noisy. In extreme cases of neglect, the pilot bearing between the constant pinion shaft and main shaft will break up, warning its neglectful owner with vibration through the gear lever, followed eventually by nasty and expensive noises.

Overdrive units are fairly tough, but need to have their oil level checked carefully. Slow engagement after fast cornering will signal low oil level and in extreme cases the overdrive may disengage of its own accord in the middle of hard cornering. Mangled overdrive internals are not uncommon, unfortunately, and the one-way clutch is susceptible to failure if already worn. Its breakage usually means replacement of the entire rear casing of the overdrive unit. Uprated overdrive parts can now be bought, which is reassuring.

The rear axles are very strong. Half-shaft breakage on an XK is exceptional, although they do bend under very hard use, particularly in racing conditions. Extreme cases can affect the brakes on a

Originality, Buying, Driving and Owning

Two gearbox layshafts. These items can cost more than £1,000 each if replacements are to be made. The one on the right has built-up gears while the one on the left is a one-piece shaft.

The Salisbury differential unit is very strong. This is from an independent rear suspension Jaguar, but the internals are the same.

Originality, Buying, Driving and Owning

Several XK axles. In the foreground is an early ENV type, easily distinguished from the Salisbury unit by the construction of the differential casing, or, if fitted to the car, by the absence of a large nut in the centre of the hub.

disc-braked car, as run-out of the disc can exceed the acceptable limits. The clutches in the Powr-Lok differential do not last too long on the non-independent rear suspension Jaguars although quite noisy clonks when turning on a tight steering lock do not always mean imminent expense. This 'clutch shuffle' is often caused by using the wrong oil, since the limited slip differential requires a different grade.

The chassis on an XK is largely rust-free, except right near the end of the frame, between the axle and rear bumpers. Checking this area is not easy, but signs of corrosion or bodging can be seen with the rear wheel removed. If suspect, the rear body will have to come off. Perhaps a better idea is to pass up such a car altogether, unless the price justifies the effort involved in rectification.

Looks aside, the bodywork on an XK is simply not worthy of the rest of the car. Rust is the main problem and the ravages of the dreaded steel weevil made many an XK into scrap yard fodder in the past. Inside the wings, along the door sills and all the hidden areas towards the rear of the car are its favourite haunts. XK 120s often suffered such bad rust to the boot floor that it would eventually rot through at the sides, allowing the spare wheel to scrape along the ground.

Originality, Buying, Driving and Owning

The body mounting outriggers of an XK 120/140. The XK 150 has three on each side. Anyone who carries out their own rebuild will curse these things as they seem to be designed to savage one's shins.

The most likely place to find chassis rot: from the end of the chassis to around 500mm forwards seems to be the most vulnerable area.

Originality, Buying, Driving and Owning

Behind the seats at the forward ends of the wheel arches is a very popular place for corrosion. This XK 120 fixed-head coupé has been treated to some very neat repair work in this region, although its battery boxes have been moved elsewhere.

BUYING AN XK

Most XKs for sale nowadays have been comprehensively rebuilt and so rust like this is not usually a problem. But even today, a shiny XK for sale can sometimes turn out to be a can of worms on closer inspection. Any bubbles at all in the paintwork, particularly around the sidelights and at the rear of the front wings, are a cue to make your excuses politely and leave. Blemishes on the surface hide far worse problems underneath.

If you are buying a cheaper example of an XK to rebuild, perhaps you are undaunted by corrosion and have the necessary metal-forming and welding equipment to deal with it; if someone will be restoring it for you, take them with you before you buy. Failing to do this could cost you thousands of pounds more than your intended budget. Nowadays, it is actually cheaper to buy a fully restored XK anyway.

Another body problem is door drop. Again, this should be rare nowadays, but the problem was a lack of greasing points in the hinges. The absence of lubrication meant that the pin would seize in the hinge and the relatively great leverage exerted by the door would mean that the seized pin was now being turned by the hinge and would gradually wear away the hinge plate. This continued until the holes became enlarged, allowing the pin to slop about and the doors to move up and down and even rattle annoyingly on poor road surfaces. This would happen on the XK 120 roadster with its small, aluminium doors,

Originality, Buying, Driving and Owning

Beautifully rebuilt battery boxes. It is wise to use one twelve volt battery as two always seem to give more trouble. The other box can then be used for storage.

The inside of an XK 140 door. The diagonal bracing will not stop the door drop, which occurs in the hinges.

Originality, Buying, Driving and Owning

but the same hinges were used throughout the range. They were asked to cope with the larger, steel doors of the later XKs, with their brass-framed windows and winding mechanisms; that they were absolutely not up to this assignment is no surprise. If door drop is encountered it can be expensive to rectify, due to difficulty of access. Any car for sale that has not had its door drop rectified has not been rebuilt properly and most likely has other problems lurking beneath its otherwise well-polished skin.

When checking any car over, waggle the steering wheel with the car stationary. On an XK 120, no play should be detected at the wheel rim, although adjustment of the nut on the steering box can remove a small amount. If there is any free play and it has not been adjusted out, how many other things has the seller omitted to do? Steering joints that are worn can be replaced, of course.

On the rack and pinion steering systems of the XK 140 and 150, free play will be apparent when applying the usual test. However, this should be only the rubber flexing on the steel and rubber rack mountings, and will disappear once on the move. However, the rubber being the sandwich filling and the two plates of steel being the bread, the rubber can tear right off one of the plates, leaving the mounting in two pieces. When in such a condition, a clonking noise will be heard as the rack waves around at one end. New mountings are not expensive, but the radiator will have to come out to replace the one under the pinion housing. (Incidentally, these mountings were drawn the wrong way around in Jaguar's workshop manual supplement.)

Another problem concerning play in the rack and pinion set-up is caused by wear in the rack bushes, usually the one opposite the pinion housing. There is no grease nipple at this end and obviously wear occurs. This can be detected by grasping the track rod around the rubber gaiter and trying to move it to and fro. A thorough dismantling will have to take place to replace a worn bush and a second grease nipple can at least be added while the rack is on the workbench.

Front suspension may need new rubber bushes, particularly those on the top wishbones. Ball joints often need replacing where they have not been greased regularly. A good check for wear in the ball joints is to lift a front wheel off the ground by jacking under the lower wishbone lever. With a lever under the tyre, any free vertical movement revealed in the top or bottom ball joints means they need replacing.

Although some cars had a composite material in the cup that supposedly did not need greasing, it is rather like the early attempts by all manufacturers to offer 'sealed for life' components. These seemed to be merely parts that cannot be lubricated and that wear out so that new ones have to be bought. The balls and sockets on the XK's front suspension would certainly outlast the car's owner if greased regularly.

As for the interior of the car, its condition is entirely visible and any improvements in condition are obvious. However, dashboards suffered quite a lot in the 1970s from owners putting in non-original parts or extra switches. There is a general code for the colour schemes in cars with two-tone leather interiors, although the factory would supply almost any colour scheme to special order.

As for current prices, it is not worth trying to give an accurate guide in a book such as this. It is interesting, however, to note the behaviour of XK values since the car's appearance. Gradually declining in price since their introduction, the XK hit its rock bottom in 1967 or 1968, with the 120 and

Originality, Buying, Driving and Owning

140 the cheapest. The XK 150, naturally enough, was the most expensive, but because it was in some cases only a seven-year-old car, it seldom sold for less than £300, which was not in those days anything like the bargain that the 120 and 140 could be. There was a marked difference between an XK 120 and XK 150 owner in those days, too.

At that time, many sought a hairy-chested sports car and the choice came down to a big Healey or an XK. A little bit of homework revealed that the XK was stronger, more practical and had a vastly superior engine. It was also cheaper, for an XK at that time could be had in running condition for well under £100. Nobody seemed to want old XKs at the end of the 1960s. Values, however, gradually crept up throughout the 1970s and many owners sold their XKs and raised a useful sum.

It was not until the late 1980s that XK prices, driven up by a dangerous combination of greed and ignorance, entered cloud-cuckoo land. The ethereal and ephemeral demand that created these lunatic prices

An XK 140 drop-head, as purchased in 1967 for £55. Those were the days.

Originality, Buying, Driving and Owning

HBC 226 was bought in 1968 for £50, but I believe the current owner knows nothing of this picture – yet!

This picture was taken in the mid-1970s, although I was not allowed to know where. Clearly someone had seen into the future.

Originality, Buying, Driving and Owning

did not just apply to XKs of course, but all desirable cars and some not so desirable, such as the MGB, which is only a cross-dressing Austin A60 after all. Many an owner made an enormous profit from their XKs as an equal number of punters readily parted with far too much money. At auction in Germany in 1989, £104,000 was paid for a 1950 alloy XK 120, but even this may be better value than the £95,000 paid for an XK 140 drop-head coupé. (Both these prices are trifling when compared to the £1.2 million paid in 1990 for the 1963 Le Mans E-type, 5115 WK, perhaps now worth £1 million less.)

After the crash, prices have settled down to much more realistic levels, even if some hopefuls advertising their cars have apparently yet to realize that the crash has happened. Around £30,000 should at the time of writing buy an XK in absolutely first class condition. Spare a thought for the seller, who very probably paid more than this for the car's restoration alone.

XKs with a history, and this nearly always means an alloy XK 120, could fetch twice or even three times this figure. With NUB 120 not for sale and HKV 455 and HKV 500 broken up by the factory (the most awesome acts of vandalism ever perpetrated, in the eyes of a Jaguar lover), only the JWK cars are really in the five-figure bracket. At the other end of the scale, a very rusty fixed-head 140 or 150 might be as little as £5,000, although such cars are not bargains, as explained earlier.

PARTS, CLUBS AND SPECIALISTS

Just about any part at all can be obtained for an XK nowadays, although the price asked will be a reflection of the part's

All manner of XK parts are being re-manufactured. These are air cleaners made by XK Engineering.

Originality, Buying, Driving and Owning

scarcity, just as the main law of economics tells us.

So where should you go to buy bits for your XK, have it overhauled, restored or made to perform better? There are a considerable number of places that advertise such services, of no doubt varying degrees of expertise, but all willing to relieve the XK owner of his money. If you are new to XK ownership, how should you find places that deal in XKs and then decide which to choose?

The first way may be to buy the bi-monthly journal *Jaguar World*. In this excellent periodical you will find an amazing number of advertisements by establishments offering restoration, engine and transmission overhaul and spares of all kinds. There is also a Jaguar classified section, often containing XKs. That the magazine is edited by Paul Skilleter should on its own be a good reason to buy it. However, how do you decide which company to do business with? Apart from considerations of geography, the best way to find out about those advertising their wares is to join a Jaguar club. There are two main clubs currently in existence, the Jaguar Drivers' Club and the Jaguar Enthusiasts' Club.

This way you can either learn from the correspondence pages and general chat in the monthly magazine, or attend one of the Jaguar meetings, usually a local one each month or a few much larger national gatherings that are usually held on a yearly basis. By talking to members, you can learn where to go – or where not to go – for the services required.

The older of these two clubs is the Jaguar Drivers' Club (JDC), which was formed in 1956 by Raymond Playford, who owned LVC 345, the two-tone XK 120 that the factory had loaned to Stirling Moss. There was an SS Car Club that was started in 1934, but it seems the secretary made off with the funds after the club's first big event in the summer of 1935. Lyons refused to have much to do with the club after that and although it continued for a while, it did not resurface after the war.

With a claimed 12,000 members, the JDC has all the normal benefits of a club of this size, as well as support from the factory, to which pilgrimages are often organized. Many other benefits are available to members, such as discounts on Jaguar car insurance, technical advice, help with finding parts and Jaguar accessories such as T-shirts and sports bags. The club also encourages the local meetings, with around seventy separate areas dotted around the country, each with its own representative. The JDC has six separate registers to cater for owners of different Jaguar models, including an XK register, co-founded by Paul Skilleter. There is also a much improved monthly magazine, *Jaguar Driver*.

For a long while the only Jaguar club, the JDC seemed intent on self-destruction a few years ago and many key members left. The antics of just a few eccentrics, however, should not affect one's opinion of the club, which is now almost forty years old. Those from the JDC whom the author contacted while researching this book could not have been more helpful. In character, the JDC seems to lean towards the social, with its emphasis on regular area meetings, motor-sport events, concours competitions and the many national gatherings organized throughout the year. The headquarters in Luton has full-time staff who can be contacted on 01582 419332, fax 01582 455412.

The Jaguar Enthusiasts' Club (JEC) is almost thirty years newer than its rival. Its membership has been growing strongly, with a certain amount of encouragement from the troubles at the JDC as well as

Originality, Buying, Driving and Owning

from being voted 'Car Club of the Year' in 1993. Currently it has 12,000 members, the same size as the JDC, but the classified advertisement pages in its monthly *Jaguar Enthusiast* (voted 'Club Magazine of the Year' three times) contain more than twice as many entries as those in the JDC counterpart. This seems odd if the two clubs have the same size of membership. Perhaps JDC members are happier with their cars and do not want to part with them?

Certainly the interest in the newer club is more on the technical side, with emphasis on maintenance and restoration – things to keep the cars running and enjoyed as they should be. With the other benefits the same as those available to the JDC members, such as technical advice, accessories and discount insurance, the JEC also has over sixty local regions and organizes national events as well. It also holds national Jaguar 'spares days' in conjunction with the more recently formed Jaguar Car Club, offering a comprehensive range of general tools, as well as special Jaguar ones, and the re-manufacture of obsolete parts. The JEC also has a book service guide that lists a large number of Jaguar books, handbooks and manuals still available. Most importantly, perhaps, it offers an excellent specialist and service directory, which lists over 500 Jaguar parts and servicing specialists throughout Britain. There is a growing list of overseas suppliers as well. The JEC is a must if you intend carrying out any of your own maintenance and repairs. It can be contacted by writing to Freepost (BS 6705), Patchway, Bristol, BS12 6BR.

There are many places to go to if your XK Jaguar needs attention – XK owners really are spoiled for choice, as a look through the pages of *Jaguar World* or the two club magazines will reveal. The examples that follow cover spares, restoration and, for the enthusiastic driver, improving the handling and increasing the power. These are just leading examples from the large choice available.

For XK spares, Jeremy Broad is both very reasonable and pleasant to deal with. His son Guy has now given his name to the business and they have between them enormous experience of XKs, apparently matched by the vast quantity of second-hand (not re-manufactured) spares, as well as new ones. Guy Broad is based just outside Coventry and can be contacted on 01676 541980, fax 01676 542427, mobile 0860 505619.

For all aspects of restoration, XK Engineering would be hard to beat, as perhaps their name implies. They undertake all bodywork, trimming, engine, transmission and chassis work. They also supply re-manufactured and reconditioned XK parts that they claim are from the largest stocks to be found anywhere. The way they have built up their premises into a first class facility is testimony to the quality of their operation. They are also situated on the outskirts of Coventry and can be contacted on 01203 622288, fax 01203 619323. For parts fax 01203 619281.

For improvements to the handling and even outright roadholding of an XK without disfiguring the car, Harvey Bailey Engineering must be one of the very best places to go. Rhoddy Harvey-Bailey himself has won many races in his famous 120 that had many clever tweaks but still looked less molested than some of the cars he led home. He also raced E-types with considerable success as well. Not many individuals know more about improving the dynamic behaviour of an XK – or any car for that matter – than Rhoddy Harvey-Bailey himself. Much has been learned in the art of suspension behaviour since the late 1960s

Originality, Buying, Driving and Owning

This shot shows both the sensible way the rear bumpers are mounted on an XK 140 and the quality of the work undertaken by XK Engineering ...

... as does this shot of the front bumper irons and how they attach to the chassis.

Originality, Buying, Driving and Owning

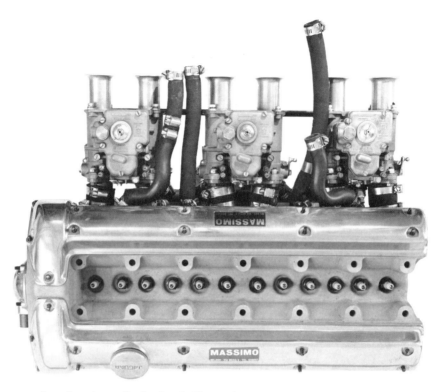

A very unusual twelve-plug cylinder head. The carburettors are Weber 42 DCOEs, not the usual 45s.

and Harvey Bailey Engineering can supply handling kits for the XK that will not affect the car's ride comfort. They also carry out a lot of consultancy work for the motor industry. They are based in Derbyshire and can be contacted on 01335 346419, fax 01335 346440.

All XK body panels, whether external (wings, doors and bonnets) or internal (bulkheads, scuttles and boot floors), are the speciality of Classic Parts and Panels. They keep a large stock of these items, but can manufacture to short notice if necessary. They also carry out restorations. Their premises are in Aylesbury, telephone 01296 398300, fax 01296 397737.

Appendix I: Production History

PRODUCTION TOTALS

The following three tables show the total numbers of XK Jaguars produced, by model, series and body style. It can be seen that the XK 120 open two-seater is by far the most common of the nine models.

By Model

	Right hand drive	Left hand drive	All
XK 120 drop-head coupé	294	1,471	1,765
XK 120 fixed-head coupé	194	2,484	2,678
XK 120 open two-seater	1,175	6,437	7,612
XK 140 drop-head coupé	479	2,310	2,789
XK 140 fixed-head coupé	843	1,965	2,808
XK 140 open two-seater	73	3,281	3,354
XK 150 drop-head coupé	662	2,009	2,671
XK 150 fixed-head coupé	1,368	3,094	4,462
XK 150 open two-seater	92	2,173	2,265
Totals	5,180	25,224	30,404

It may also be of interest to compare each model in terms of ascending order of sales:

Right-hand drive			Left-hand drive			All		
140	OTS	73	120	DHC	1471	120	DHC	1765
150	OTS	92	140	FHC	1965	150	OTS	2265
120	FHC	194	150	DHC	2009	150	DHC	2671
120	DHC	294	150	OTS	2173	120	FHC	2678
140	DHC	479	140	DHC	2310	140	DHC	2789
150	DHC	662	120	FHC	2484	140	FHC	2808
140	FHC	843	150	FHC	3094	140	OTS	3354
120	OTS	1175	140	OTS	3281	150	FHC	4462
150	FHC	1368	120	OTS	6437	120	OTS	7612

Production History

By Series

	Right-hand drive	Left-hand drive	All
XK 120	1663	10,392	12,055
XK 140	1395	7556	8951
XK 150	2122	7276	9398

By Body Style

	Right-hand drive	Left-hand drive	All
Open two-seater	1340	11,891	13,231
Drop-head coupé	1435	5790	7225
Fixed-head coupé	2405	7543	9948

PRODUCTION LIFE SPANS

When viewing these figures it is important to note that deliveries would continue after production ceased; also that precise totals are not possible owing to slight discrepancies between Jaguar's chassis records and delivery inventories.

	XK 120	XK 140	XK 150
Production started	July 1949	October 1954	May 1957
Production ended	September 1954	February 1957	January 1961
Length in production	58 months	30 months	44 months
Number made	12,055	8,951	9,398
Approximate monthly rate	210	300	210

PRODUCTION HISTORY – XK 120

Introductions

Open two-seater: 22 October 1948 (no real production until July 1949)
Fixed-head coupé: 2 March 1951
Drop-head coupé: 18 April 1953
Special Equipment: September 1952

Chassis Numbers

The chassis number on an XK is stamped on the top surface of the chassis next to the exhaust downpipes. On an XK 120 it is also supposed to be stamped three feet forward of this point, on the front of the left-hand chassis side member, although the author has never found one there. Changes took place from and including the chassis numbers given, although occasionally there would be a small number of 'stray' cars where the update in question took place one or two chassis numbers before or after the main changes.

	Right-hand drive	Left-hand drive
Open two-seater	660001	670001
Fixed-head coupé	669001	679001
Drop-head coupé	667001	677001

'S' Prefix indicates Special Equipment model

Engine Numbers

From W1001. After W9999, from F1001 (November 1953).

Visual Production Changes

1949/50
660019/670027 – Curved windscreen side pillars
660029/670069 – Starting handle deleted

April 1950
660059/670185 – Steel body replaces alloy; longer stem for interior mirror; passenger grab handle now mounted on fascia; headlamp nacelles longer; windscreen pillar grommets smaller

November 1951
660675/671097 – Footwell ventilators in front wings
660729/671098 – Hood lengthened, now fastened further back
660911/671493 – Heater standard

February 1952
660935/671797 and 669003/679222 – Salisbury or ENV axle now fitted

April 1952
660980/672049 and 669003/679622 – Tandem master cylinder with dual reservoir

May 1952
Engine W5465 – 6-blade fan replaces 4-blade cast alloy type

September 1952
Dashboard layout changes (*see* notes)

October 1952
Engine W6149 – Oil level element and indicator deleted

Production History

Screenwashers are now a legal requirement. This is a reservoir as fitted to an XK 150.

661025/672927 – Sidelights integral with wing
661026/672963 – Windscreen demister vents

December 1952
661037/673009 – Screen washers

March 1953
669005/680738 – Reverts to single exhaust pipe for SE model

April 1953
Salisbury axle only now fitted

September 1954
661172/675926, 669194/681481 and 667280/678418 – Flat horn push (as the XK 140)

Colours

It is outside the scope of this book to try to list all the colours that were offered by the factory for each of the three XK types, because the standard colours listed varied each year. Furthermore, just to make the task harder for a chronicler of XK hues, customers could order almost any colour scheme they desired, outside the standard range.

The engine bay on early XK 120s was black; later engine compartments were painted in the body colour. The underneath of the wings was flat black, not rough-textured like underseal. The underside of the bonnet was in the body colour. The chassis was flat matt black (not gloss or rough finished), as also were the rear axle, petrol tank, brake drums and suspension – almost the whole rolling chassis.

Exteriors The following colours were available for the exterior of the XK 120 as full-scale steel production got under way in 1950: birch grey, red, suede green, silver, pastel blue, pastel green, cream, black, bronze.

Interiors Where two-tone combinations were employed, dashboard and seat flutes were the lighter colour; beading along bottom of dashboard, carpets and door trims the darker one. Later open two-seater cars had one-colour interiors as standard, as did all fixed-head coupés and drop-head coupés. Customers could, however, specify two-tone interiors to special order. Colours available were: suede green, red, biscuit, tan, pale blue, grey, French grey, black, dark sand, gunmetal.

Hoods Black, French grey, gunmetal, fawn.

General Bumper irons were black, or body colour on some lighter-coloured cars. Solid wheels were body colour. Wire wheels were also body colour, but later cars had a chromium plating option; this sometimes excluded the rim, which was painted silver. Hubcaps (also known as 'nave plates') were painted in the body colour between the rim and the centre piece. Hubcaps were slightly shallower on the early cars. A mistake nowadays that occurs regularly on all non-black XK models is placing black beading between the rear wings and body. The rear wing beading should be body colour, so unless the car wears black coachwork, black beading is wrong.

Chromium plating Bonnet and boot lid props on early cars were plated, as were all bumper dome nuts, spacers and washers. Also plated were all windscreen fixings on open two-seaters, the reversing light and tail lights. The separate sidelights on early open two-seaters were also chromium-plated.

Notes

On early cars, the starting carburettor was attached to the rear carburettor, on later cars to the front. Early cars also had long plug leads that ran from the distributor under the carburettors and into the head from the rear. This tended to encourage trouble with arcing, so they were shortened to climb over the head at the front, a route retained until the end of XK engine production.

From mid-1952, camshaft covers had eleven studs each, increased from eight. The changeover point is rather blurred and therefore impossible to determine exactly.

In August 1954 legislation was introduced to make two rear reflectors compulsory. These should be fitted 3in (75mm) from the bottom and 1.5in (40mm) from the edge of the boot lid.

The most common mistake on the XK 120 used to be fitting the rear overriders upside down; the fixed-head coupé supplied by the factory to *The Autocar* to road test had them thus fitted. Although this visual inaccuracy is rare nowadays, it may be worth noting that they should be fitted fatter part uppermost.

The interior assembly, rim and glass in the early chrome sidelights were the same as on a Mk II Jaguar, except that the XK lens carried no lettering.

Drop-head and later fixed-head seat backs were much flatter than on the roadster and early fixed-head coupé models. While roadster seats had eleven flutes on the cushion, other XK 120s had ten.

The instrument panel on the roadster changed after September 1952, to incorporate a rheostat heater switch and an extra warning light for the trafficators. Also changed was the

Production History

lighting switch and gone were the oil-level push button with the dual oil level and petrol gauge. The panel was also slightly rearranged with, for example, the ignition key at the bottom where other keys on the key ring could not wear away the leather. This new layout would be common to both the XK 120 coupés, apart from a few early fixed-head models, and was continued on the XK 140.

The Lucas number plate lamp was also fitted to some Triumph Vitesse models, the door-locking knob ('handle for safety catch') on the open two-seater was also fitted to Sherpa-type vans, and the boot lid handle was also fitted to Bedford Dormobiles.

XK 120 number plate lamp.

PRODUCTION HISTORY – XK140

Introduction
All models: 15 October 1954

Chassis Numbers

	Right-hand drive	Left-hand drive
Open two-seater	800001	810001
Fixed-head coupé	804001	814001
Drop-head coupé	807001	817001

'A' prefix indicates Special Equipment model
'S' prefix indicates Special Equipment, with C-type head
'DN' suffix denotes overdrive gearbox
'BW' suffix denotes automatic transmission (Borg Warner)
'CR' suffix on gearbox indicates close ratio gears

Visual Production Changes

September 1956
800072/812647, 804767/815755 and 807441/818729 – Fly-off handbrake

October 1956
804781/815773 and 807447/818796 – Steel doors (not on open two-seater)

Colours

Exteriors Black, Pacific blue, red, British Racing Green, cream, maroon, suede green, birch grey, pastel green, pearl grey, dove grey, lavender grey, pastel blue.
Interiors Red, tan, grey, blue, suede green, light blue, biscuit. Two-tone to special order.
Hoods Black, French grey, fawn, blue, sand, gunmetal.

Notes

The steering wheel was 17in (430mm) in diameter, a reduction of 1in (25mm) over the XK 120. XK 140 seats had nine flutes on the seat cushion. The instrument panel on the open two-seater had curved sides. The open two-seater screen was the same as on the corresponding XK 120.

XK 140 hubcaps were chrome finish only. Two styles of radiator header tank were fitted – on early cars they had flat tops, later they were fluted and looked less substantial. The ends of the exhaust tail pipes were chromed on all models.

PRODUCTION HISTORY – XK 150

Introductions
Fixed-head coupé: 22 May 1957
Drop-head coupé: 22 May 1957
Open two-seater: 28 March 1958
'S' open two-seater: 21 March 1958
'S' fixed-head coupé: 18 February 1959
'S' drop-head coupé: 18 February 1959
'S' 3.8: 2 October 1959

Chassis Numbers

	Right-hand drive	Left-hand drive
Open two-seater	820001	830001
Fixed-head coupé	824001	834001
Drop-head coupé	827001	837001

'DN' suffix denotes overdrive gearbox
'BW' suffix denotes automatic transmission (Borg Warner)

Production History

'CR' suffix on gearbox indicates close ratio gears

Visual Production Changes

May 1958
824523/835301 and 827011/837332 – Arm rest doubles as door pull

June 1958
824414/835548 and 827069/837415 – 60-spoke wire wheels replace 54-spoke; indicator operation lever now on steering wheel

July 1958
830439, 824420/835566 and 827072/837434 – Rheostat heater switch next to rev counter (no right-hand drive roadsters were affected by this change)

April 1959
820001/831250 and 827209/837662 – Rear overriders moved closer together

May 1959
820004/831712, 824669/835886 and 827236/837836 – Square brake pads

June 1959
820001/830001, 824677/835893 and 827240/837865 – Sprung boot lid replaces strut
820014/831825, 824702/835905 and 827258/837865 – Ashtray moved from door to transmission tunnel
820017/831899, 824742/835935 and 827272/837941 – Brake vacuum tank fitted
820039/832076, 824864/836187 and 827355/838246 – One air cleaner replaces three wire ones ('S' models)

August 1959
820071/832120, 825179/836744 and 827540/838754 – Brake fluid warning light and level indicator

Colours

Exteriors Pearl grey, mist grey, Cornish grey, British Racing Green, Sherwood green, indigo blue, Cotswold blue, Carmen red, claret, maroon, black, cream.
Interiors Red, maroon, suede green, light blue, dark blue, grey, tan.
Hoods French grey, blue, sand, fawn, gunmetal, black.

Notes

XK 150 discs are the same front and rear. A Mk I 3.4 grille is slightly different from the grille on an XK 150, but will fit.

PERFORMANCE

The following figures are intended to be a guide as to the *likely* performance of the various XKs. Performance figures for apparently identical models are affected by so many things. Engines from the same production line yield slightly different horsepower outputs, different air temperatures and densities affect outputs, and different driving techniques produce different results even for the same car. Tyre pressures, wind direction, inclination of road used and state of tune of the car will all have an effect. It therefore does not make a lot of sense to give the elapsed times to any greater accuracy than half a second. In particular, the top speed figures should not be taken as icons. The true top speed of two apparently identical cars in perfect tune can be expected to vary by two or three miles per hour (3 to 5km/h) either way. Where possible, the figures given are the mean of those available from contemporary reports.

The last column in the table gives figures for a XK 120 with a 3.8 XK150S engine or equivalent: 3,781cc with triple carburettors and straight-port head.

It can be seen that the XK 140 and XK 150, being overdrive models, have usefully improved touring fuel consumption.

Specification	XK120 Standard	XK120 SE	XK140 C-type	XK150 SE	XK150 S	XK150	XK120 3.8S
Max gross bhp	160	180	210	210	250	265	265
Max torque (lb/ft)	195 at 2,500rpm	203 at 4,000rpm	213 at 4,000rpm	216 at 3,000rpm	240 at 4,500rpm	260 at 4,000rpm	260 at 4,000rpm
0–60mph (sec)	12	10	11	8.5	8	7.5	6.5
0–100mph (sec)	35	28	30	25	21	19	17
30–70mph (sec)	11.5	10.5	11	8.5	7.5	7	6.5
Max speed (mph)	117	120	127	125	132	136	140
Max speed (km/h)	188	193	215	201	212	219	225
Fuel consumption (mpg)	14–19	14–18	17–23	16–23	15–22	14–21	–
Fuel consumption (litres per 100km)	15–20	16–20	12–17	12–18	13–19	13–20	–

Appendix II: Contemporary Road Reports

> **ROAD TEST**
> XK120 Jaguar Super Sports
> Reproduced from *The Autocar*
> 14 April 1950

'No, it's not a racing car,' was an answer that had to be given several times to small boy admirers of the Jaguar XK120 while it was with *The Autocar* for Road Test. Perhaps there are others who do not appreciate that this stupendous car of the sleek appearance is primarily a very fast, tractable touring car and not 'a racer,' even though examples of the model have appeared with great success in sports car events, notably the Production Car Race at Silverstone last August.

Fresh in mind, too, will be that remarkable performance on the Belgian motor road in 1949, when one of these cars, running with an undershield, which is optional equipment, and fitted with a racing type windscreen, which, again, is available, achieved 132.6mph over a flying mile and 126mph with normal windscreen and hood and side screens erected, thus making the XK120 indisputably the fastest series production car in the world. Owing to the virtual impossibility of attaining such speeds in England at present, ultimate possibilities as regards maximum speed have been taken as read, in view of that officially certificated performance, which was witnessed by a member of the technical staff of this journal; but readings up to 117 have been seen on this present occasion on a speedometer only 4 per cent fast at an indicated 100.

In trying to convey in a word-picture the supreme position which the XK120 two-seater occupies, there is a temptation to draw from the motoring vocabulary every adjective in the superlative concerning the performance, and to call upon the devices of italics and even the capital letter!

It has a power-to-weight ratio which gives it the heels of any car produced in series; better than 122bhp per ton is an extraordinary figure for a production car. There is the astonishing fact to keep in mind that it is listed at the same home market price as the 3½-litre Jaguar saloon, though unfortunately the home enthusiast is all but barred from buying the XK at present. Whilst a full 100mph can be treated as a timed acceleration test, and only sufficient breathing space is needed to see 110, 115mph, and more, on the regular production car as now issuing from the Coventry factory is some numbers, it is also remarkably docile and capable of being driven on top gear at 10mph.

Nothing like the XK120, and at its price, has been previously achieved – a car of tremendous performance and yet displaying the flexibility, and even the silkiness and smoothness of a mild-mannered saloon.

The heart of this astonishing versatility is a 3½-litre six-cylinder twin overhead camshaft engine that develops 160bhp at 5,100rpm with astonishing smoothness, and maintains that figure on a flat peak of the power curve towards the 6,000rpm mark. This power unit is a British achievement in which everyone in this country interested in cars of high performance may well take pride. Indeed the XK is a prestige gainer for Britain's engineering as a whole and car engineering in particular. During a test of some 700 miles, at the beginning of which it was brand new and by no means run-in, it necessarily received some merciless treatment, but showed no sign of losing tune, used very little oil and did not at any time record above 80°C water temperature.

More than usually, study can be commended of the performance figures in the accompanying data panel. These show an ability to reach very high speed in phenomenally short spaces of time (notably from rest through the gears to a genuine 100mph in not much more than half a minute – 33.5 sec was the best recording), and also top gear acceleration, even from speeds as low as 10 and 20mph, of an order associated with the best of biggest engined saloons designed for top gear performance. Yet the XK 120 runs on a top gear ratio as high as 3.64 to 1; 3.27 to 1 is available optionally.

Dual Character

Truly this is two cars in one. It can be handled quietly with very little use of the gears if the driver is in a lazy mood. Press the right foot hard down, however, and a different car is revealed. A snarl comes into the exhaust note, though never excessive noise, and on a familiar road the bends and even the landmarks seem to have been redesigned overnight, and placed much closer together than had previously been realized!

Still further illustration of the top gear powers was provided on a 1 in 9-maximum hill, which has one bend of nearly 90 degrees, and which usually requires third and second on other cars and occasionally is climbed on top gear. This the XK scaled at 40mph on top until it was baulked on a blind bend.

The usual 1 in 6½ hill of these tests became almost a Shelsley Walsh affair – after dark, for the sake of the greater safety factor implied with head lights in use.

Probably with rushing tactics even this gradient could have been scaled on top gear, certainly fast on third, but deliberately second was used to give maximum performance on a hill which is far from straight. The extraordinary rate of 58mph by corrected reading was reached on this quite severe hill, where to climb at even 35–40mph is exceptional. The driver concerned at this stage, familiar with this hill over the past twenty years in most of the world's makes and models, feels that he will need to live long before he improves on that performance, or, indeed, on most of the recordings put up by the XK during this test.

Even town traffic is a pleasure in it, because the driver can nip in and out swiftly without ever becoming a nuisance to others because of noise. Minutes can be clipped off customary times even for short journeys, whilst on longer runs the average speed is limited only by the enterprise of the driver and the traffic encountered. The cruising speed is as fast as you can drive it. Given the right conditions, the average speed of a lifetime could obviously be achieved – and the driver rest on his laurels for evermore.

The control is exactly as one would wish with such a car – firm but light steering, highish geared, with sufficient caster action, allowing of a quick swerve with complete safety; and suspension which ties the car down and yet is thoroughly comfortable. It is well damped by telescopic hydraulic shock absorbers for the torsion bar independent front suspension, and big hydraulic dampers for the half-elliptics at the rear. With normal tyre pressures of 25lb per sq in the riding is in every way comparable with that of the best independently sprung cars of today.

Yet the XK can be hurled round bends with quickly increasing confidence, and, as was particularly shown by a few fast laps of an unofficial circuit on a disused airfield, extremely fast cornering is achieved with no more than tyre scream and the body leaning over so far but no farther, so to speak, in a way which does not cramp the driver's style. For the high-speed part of the test the Dunlop Road Speed tyres were inflated to 35lb per sq in, as recommended for sustained faster work. The riding is then harder, as would be expected, but still is not harsh.

Brakes and Gear Change

The Lockheed hydraulically operated brakes are given a tough task on such a car, but with the special linings used on this model they did not fade and at all times the driver felt that the speed was under control, whilst the brakes could be used hard with safety at high speed. The central gear change is by a short, rigid lever which is placed rather too far back to be ideal at a close driving position, though the movements are pleasing. The synchromesh for second, third and top does not intrude. Very fast downward changing is achieved without beating the synchromesh, by employing the old double-declutching technique, or more leisurely changing is made smoothly and quietly, taking full advantage of the synchromesh. Third gear is silent and can be used with tremendous effect for alternate deceleration and acceleration on a winding road.

The car that has been handled in succession by *The Autocar*'s 'high-speed flight,' each member of which has unstinted praise for the performance, represented the first of the production examples with a steel body. Experience has been had also of one of the earlier aluminium-bodied cars that ran at Silverstone last year, the bulk of the present testing being carried out on the production car. Independent weighing showed, contrary to expectation, the steel-bodied car to be some 40lb *lighter* than the early example, in the running trim as tested.

This production car had left-hand drive as for the USA. and elsewhere, and the high compression ratio of 8 to 1 intended for high octane fuel available in some countries. Part of the test, including recording the performance figures, was carried out on 80-octane petrol, as available to the factory for test work. On 80-octane there was practically no pinking, and then only at low speed, and even on normal Pool petrol (approximately 70-octane) the pinking was not violent and running on did not occur.

A first-press start from cold was obtained on every occasion, and without sign of temperament or need for warming up, as used to be expected of a really high-performance engine.

So much needs to be said of this car of cars on the road that little detail can be given regarding the bodywork and equipment. Elbow room is adequate, and the separately adjustable front seats, upholstered in fine-quality leather, give suitably upright positions, and good support to the shoulders. Driving vision is good, with a very satisfactory view of both wings, although a short to medium-height driver would prefer the top of the telescopically adjustable steering wheel to be lower.

The mirror view is good. There is useful luggage space and good provision for carrying oddments is made in wide pockets in the thickness of the doors. The head lamp beam is useful up to, say, 90mph on known roads.

ROAD TEST
Jaguar XK150S

Reproduced from *The Autocar*
18 September 1959

There seems to be no end to the development of the Jaguar XK series engine. In the XK150S, as opposed to the normal 150, compression is up to 9 to 1, and three instead of two SU carburettors are used in conjunction with even straighter porting. There are other less apparent differences, including quick change pads for the four Dunlop disc brakes, and the car under review was fitted with the optionally extra Powr-Lok limited slip differential.

To realize how far this outstanding engine has come, one must remember that the first XK fully tested by *The Autocar* in 1950 had the then staggering output of 160bhp at 5,100rpm. Now the 150S gives 250bhp at 5,500. The original two-seater reached 100mph in 35.3 sec, which has now shrunk to 22.4 sec, in spite of greater size (there is room for children in the back) and an increase in weight. While the latest performance figures are fully as impressive today as those of the earliest car, judged by any standard, this Jaguar is yet a solidly built vehicle weighing, with driver, fully a ton and a half.

The car is appreciably faster than the standard XK150 currently in production. When last tested, the normal model took 25.1 sec to reach 100mph, and was 13.5mph slower at maximum speed.

This is a driver's car in all respects, yet the engine's exceptional flexibility makes it also a splendid ladies' town carriage. Regard the performance data, which, in stark printing ink, fail miserably to convey the sheer thrill of the real thing. When 136mph was achieved, even the Continental auto-route seemed to have exchanged its curves for corners. Yet the car remained rock steady. Indeed, at one stage the observer ensconced himself in the rear compartment and happily photographed the speedometer needle at its exciting limit: 140mph (136 true speed). Even the engine was turning at a modest 5,100rpm – well within its 6,000 limit.

In practice 5,500rpm on the tachometer, coinciding with the beginning of the red danger segment, is as much as can be used with benefit to the acceleration. Treating this figure as the limit, therefore, just as a private owner would, speeds on the gears were found to be 33mph in first, then 59, 86, and 111mph in top before switching in the Laycock overdrive. Something of the car's extraordinary acceleration in the high speed range can be gauged from these figures. With a top gear maximum of 111mph and 25 more mph in hand for overdrive, the acceleration upwards of 80mph is as fierce when the car is roused as that of many a good sports car pulling away from rest.

Partly owing to the superb, never-fade brakes, 90mph can be reached repeatedly on British roads *in safety*. It is at such speed, and when even more is wanted, that the acceleration is at its most exhilarating. After driving the 150S for many miles the driver realizes that he is in a class apart from ordinary traffic.

Even the traffic queues of a sunny summer Sunday lose much of their significance. A group of half a dozen vehicles or so can be overtaken in next to nothing of a straight, again in complete safety. When the traffic is at its worst, when most cars even in expert hands are fumbling their way along our main roads, the Jaguar still manages to achieve altogether exceptional average speeds.

In spite of the new power output the engine is as smooth as ever; it remains one of the most silky units in production, having turbine sweetness. Flexibility is such that (while no owner would do it, of course) the car can be made to pull away from rest without fuss in top gear. Out of interest, one run was made from a standstill to 100mph in top gear only, and the remarkably short time of 33.5 sec resulted. The suitability of the torque curve, coupled with flexibility, mean that between 20 and 100mph the car will accelerate in top at a very even rate.

Only 2.7 sec is needed to reach 30mph from rest, and 60 – about the maximum cruising speed used by most drivers of other cars – can be reached in 8.9 sec. To reach even 120mph takes fewer than 40 sec. The standing start quarter-mile time of 16.2 sec is but 0.1 outside being a record for production cars tested by this journal.

An appreciable reduction in axle hop is ensured by the limited slip differential, and wheelspin during fast starts is checked; in normal drive axle hop is noticed only when accelerating hard out of sharp corners in second gear. In this respect alone does the increased power output appear to be getting ahead of the basically unchanged chassis design.

There is nothing new about the gearbox, which has the familiar, four speed, central change coupled with the switch-operated overdrive on top. Synchromesh, provided on the upper three ratios, is not always adequate even for leisurely changes, and its weakness takes away some of the pleasure in running up quickly through the gears. The overdrive is protected from fierce jerks, the throttle having to be at least partly open before it engages or frees itself. The switch is passably well placed, though close to the indicators' control and to the tip of the ventilator window when this is open.

A stronger than standard clutch is used in the S model, and there is yet another version available for competition work. That on the test car proved quite up to hard driving. Only during prolonged traffic work in London did the pressure needed to declutch become slightly fatiguing to the left leg. There is room for the left foot away from the clutch. Smoothness of the take-up is reasonable. The throttle pedal is placed farther from the driver than the brake, which makes heel-and-toe operation difficult – for some drivers impossible.

Pleasure of handling the car was diminished by some fierceness of the throttle linkage. Considerable pressure was required on the pedal, and, particularly between 1,500 and 2,000rpm, it was insensitive when the engine was under light load. Appreciable vigilance was required to avoid jerky starts in normal motoring. With so much power available, control on icy surfaces could not be as delicate as one would like.

Many drivers, even of sports cars, use their brakes little while yet covering the ground very quickly by normal standards; deceleration under engine braking is often enough to check from 70mph or thereabouts for the bend ahead. But fast travel in this Jaguar on British roads calls repeatedly for firm braking. As such very high speeds can be reached on short, clear straights, one may say that almost every corner calls for some prior use of the brakes. With similar frequency the car catches up other traffic so quickly that speed must be reduced appreciably before the driver commits himself to overtaking. The Dunlop discs as fitted to the 150S are so good that they will take any treatment handed to them. First and foremost there is no fade – full braking power is always on tap without variation in efficiency. Next, the actual power is very high indeed, with 96 per cent efficiency available on dry concrete. Third, they stop the car in a straight line from any speed, and in conjunction with other aspects of the car's design can be put on hard at three-figure speeds without adversely affecting control. Fourth, the servo mechanism ensures that pedal pressure is very light pro rata to retardation provided.

Maximum braking power calls for only 85lb pressure, and from the gentlest of check braking to the emergency full stop the system is both smooth, sensitive and progressive. One gets the impression that the car could be loaded to the roof and raced down the highest Alp, stopping with the same calm efficiency at the bottom as at the top. A minor criticism of the brakes proper is that they produce a squeak when being used gently at low speed; this sound is often heard from disc brakes.

The hand brake lever is of the fly-off type; that is, the button in the tip has to be pressed in to secure the lever on, release being effected by pulling the lever till the button springs out, and then simply letting go. It is mounted conveniently between the seats and it is possible for the driver to exert considerable pressure on it. However, Jaguar – or Dunlop– have not overcome the difficulty of providing a hand brake which will operate strongly on disc brakes, which themselves require powerful servo assistance in normal use. The lack of power of the hand brake, in conjunction with the pedal layout, makes smooth restarts on steep hills far from easy.

A considerable variation in tyre pressures is recommended for different conditions, the car's speed range being so great. These have their effect on the ride. The suggestion for normal motoring is 23lb front and 26lb rear, raised to 30 and 35 respectively for sustained fast driving. However, if speeds near the maximum are maintained for long periods on motorways, even 10lb more all round than high speed pressures is advisable. In practice the lowest pressures, while giving a softer ride, permit considerable tyre squeal at moderate speeds, and 30–35lb sq in seems best for general use, whether or not rapid travel is envisaged. The suspension generally is good; it is not harsh, and while it permits slight roll on corners taken really fast, the car remains absolutely stable. Poor surfaces can be taken fairly quickly, but there is some patter if the speed is too high. The car behaves well in the wet, even though adhesion is appreciably reduced. In such conditions some care must be taken not to unleash too much power in the lower speed range.

One develops mixed feelings about the steering. Turns of the wheel from lock to lock at 2¾, coupled with a reasonably tight turning circle, give precise control, and for low speed manoeuvring and in normal road work the steering is always light and direct. It remains direct at high speed, but becomes appreciably heavier on fast cornering. In the front suspension layout there is little or no caster angle, self-centring being achieved instead by use of a relatively steep king-pin inclination. This means that the weight of the car is lifted around this inclined axis to obtain the degree of self-centring. On corners taken at speed, the load becomes very heavy on the outside wheel owing to weight transfer, and has its effect on the steering.

A traditional, four-spoked black wheel is provided, with Jaguar motif and horn button in the centre. There is telescopic adjustment, and any driver can quickly make himself comfortable and at home. The driving seat provides good support for one's back and thighs, but not sufficient laterally to offset centrifugal force on corners. The relationship between reach to the pedals and that to the steering wheel can be made just about ideal, with seat and wheel suitably adjusted.

Instead of being mounted in front of the driver as they should be, the instruments are central, the panel being dominated by the rev counter and speedometer dials. The needle of each dial moves clockwise as speed rises, and while the white on black lettering is clear, the driver would be much happier if the dials' positioning were reversed. That on the left (rev counter) is easy to read at any time, but the driver's hand on the wheel obscures all of the speedometer except for the top end of the high speed range. If they were mounted the other way round the speedometer could be read quite easily, and the all-important upper range of the rev counter could also be seen without difficulty.

Other gauges cover fuel level, incorporating a red warning light, water thermometer and an oil pressure indicator. There are tell-tales for main beam, indicators and ignition. Standard equipment includes heater and cigarette lighter. The wipers have two speeds, and if the central control for the lamps is turned past the main beam position, fog lamps are brought into use. In detail the coachwork and its fittings earn some praise. There is a central ashtray behind the gear lever which is well suited for knocking out pipes in addition to discarding cigarette ash. The main side windows wind right down, albeit rather slowly; there are also swivelling ventilator windows in the front doors; and the rear quarter lights are hinged at their forward edges. The trim generally is good, leather being used for the upholstery, and carpeting of good quality for the floor. However, the fluffy head lining would become discoloured quickly, and sections of the dust and draught sealing strips round the door openings were not anchored securely.

Tank capacity is 14 Imperial gallons, which gives a useable range of some 200 miles when the car is driven hard, mpg in these circumstances dropping to 16mpg With leisurely driving, which in the XK150S means pottering along at least in the sixties in appropriate circumstances, the mpg rises to 23. Oil consumption proved to be 1,670mpg, but because many miles at maximum speed were included in this figure, in addition to all the acceleration testing, it is reasonable to assume that owners would enjoy an appreciably lower demand.

Although the windscreen is shallow, visibility is satisfactory. There is less of the beetle-browed effect which the XK140 produced for many drivers, and the wide rear window gives, via the mirror, quite a good view of what has been left behind. The little seats in the rear compartment are instantly detachable should the extra space be required for luggage. With two up, accommodation for baggage is excellent. Even the shallow locker will hold a worthwhile number of suitably shaped bags. The spare wheel lies in its own compartment under the floor. Here also is a comprehensive range of tools. In the roll is an array of spanners and other hand tools, including a tyre pressure gauge, and fastened to the underside of the spare wheel cover are larger emergency aids, including the jack and grease gun. There is a small, lockable glove locker in the facia.

A good range is given by the head lamps – certainly enough for high speed to be used at night. The high frequency horns penetrate quite well, but from inside the car they sound a little unworthy of such a vehicle. Under the very crowded bonnet, dip-stick and other components needing examination or routine service are easy to get at. Manually operated struts are used for both bonnet and boot lids.

One is left with many outstanding impressions of this extraordinarily potent car. There is the racing car performance, backed by equally fine, fade-free brakes. There are the thoroughly safe handling characteristics. One cannot help being especially impressed by the flexibility of the engine when more restrained travel is the order of the day. And while the exhaust note becomes exciting at full throttle, in no circumstances could the car be called noisy – at least, by the occupants. By XK150S is a superlatively satisfying way to travel on the open road.

Index

Adams, Ronnie, 144
Aintree, 109, 119
Alfa Romeo, 89, 90–92
Allard, 28, 81, 82, 86, 92, 93, 97, 98, 101, 106, 129, 132
Allard, Sidney, 93
Alpine Trial/Rally, 13, 14, 87, 123–125, 129, 131–133, 137, 141, 142, 144
Alvis, 10, 19
Ambrose, Martin, 144
Anzani, 17
Appleyard, Ian, 88, 123–133, 136–138, 142–144
Armstrong Siddeley, 14
Arnold, Allan, 101
Ascari, 80, 89, 95, 99, 100, 102
Aston Martin, 37, 81, 86, 93, 99, 101, 143, 144
Auburn, 29
Austin, 9, 16, 140, 166
Austin Healey, 114, 166
Austin, Herbert, 9, 10
Austin Swallow, 13, 16
The Autocar, 23, 26, 28, 35, 46, 55, 149, 177, 182, 184, 186

Baily, Claude, 17, 19
Barsley, M., 124
Baudoin, 130
Beaulieu, 137
Beck, Robin, 144, 116
Bedford, 178
Bennett, Air Vice Marshall and Mrs, 138
Bentley, 15, 93
Berry, Bob, 112–113
Bidée, 130
Biondetti, Clemente, 88–91, 98–100, 116
Bira, B., 19, 80–85
Bircotes, 123
Birkin, Henry, 19
Black, Sir John, 12, 18, 19
Blackpool, 8, 11
Bleakley, William, 144
BMW, 22, 28–30, 38, 39, 63, 80, 83, 127, 133
Boddy, William, 38, 81, 100
Bolster, John, 84
Bolton, Peter, 111
Boreham Wood, 106, 110
Bornigia, Mario & Franco, 89
Boshier, S.J., 106, 110
Bourgeois, Joska, 103
Bracco, Giovanni, 89
BRDC, 95
Brescia, 91, 99, 100
Brighton Speed Trials, 79

Bristol, 17, 103
Broad, Guy, 170
Broad, Jeremy, 170
Broadhead, Jack, 101, 132, 136
Brough motorcycle, 12
Brown, Greta, 9
BTCC, 63
Bugatti, 18, 34, 106

Cadillac, 87, 93
Carraciola, Rudi, 102
Chapman, Colin, 15
Chrysler, 40
Chula, Prince , 80
Citroën, 42
Claes, Johnny, 103, 129, 130, 133, 138
Clark, Peter, 92
Classic Parts and Panels, 172
Clyno, 10
Cobb, John, 15
Cole, Tom, 93
Connaught, 79, 81
Cooper, 96
Covent Garden, 99
Crespin, 131
Crook, Anthony, 103
Crossley Motors, 8
Crowther, Martin, 118
Cystal Palace, 116
Culpan, Norman, 81–85, 102
Cunningham, Briggs, 87, 93
Cunningham, John, 142

Daily Express, 12, 35, 79, 100, 135
Daily Herald, 95
Daily Telegraph, 129
Daimler, 19, 47, 120
Datsun, 42
Decauville, 9
Denham-Cooke, G., 132
Dewis, Norman, 135, 136, 139
Dodson, Charlie, 101–103, 136
Donington Park, 95
Duesenburg-Ford, 87
Dunrod, 95, 96, 99

Earls Court, London, 28, 32, 49
East African Safari Rally, 129
East Anglian Motor Club, 127
Eastbourne Rally, 131, 132
Ecurie Ecosse, 112
England, 'Lofty', 19, 80, 88, 97, 139, 140

189

Index

ERA, 15, 80, 83
Evans, P.J., 9
Evian-Mont Blanc Rally, 138
EX135–MG special, 26
Exeter Trail, 127

Fairman, Jack, 134, 135
Fairsten, M., 113
Fairthorpe, 138
Falkenhausen, 133
Fangio, Juan, 89
Fantuzzi, 100
Ferrari, 9, 37, 73, 87, 90–96, 98, 99, 103, 114, 130, 131, 137
Fiat, 10, 99, 100
Fisher-Skinner, 121
Fitch, John, 97
Foleshill, 11, 13, 17, 18
Ford, 13, 30, 44, 75, 147
Fraikin, Charles, 137, 138
Francorchamps, 103
Frazer Nash, 30, 81–86, 93, 98, 103

Gardner, 'Goldie', 26–28, 122
Gatsonides, Maurice, 129, 133
Gatti, 99, 100
Gee, Neville, 101
Gendebien, Olivier, 137, 138
Geneva, 17, 88
Gerard, Bob, 97
Ginther, Richie, 97
Goodwood, 106–108, 110, 113, 137
Griffin, Eunice, 120
Grounds, Frank, 131, 143
Great American Mountain Rally, 144

Habisreutinger, Ruef, 124, 128, 129, 132
Hache, 131
Haddon, Eric, 142, 144
Hadley, Bert, 92, 93, 134
Haines, Nick, 87, 94–96, 99, 124, 125
Hall, Ian, 144
Haller, 91
Hally, J., 141
Hamilton, Duncan, 95, 100–102, 107, 140
Hangsen, Walt, 97
Harvey Bailey Engineering, 154, 170, 172
Harvey-Bailey, Rhoddy, 57, 88, 116, 170
Hassan, Wally, 15, 17, 19, 27, 41, 80
Hay, J., 131
Head, Michael, 107
Healey, 81, 82, 87, 93, 95, 101, 124
Healey, Donald, 87
Heath, C., 131
Henlys, 10, 133
Herzet, Jacques, 130
Heynes, William, 14, 15, 17, 19, 23, 32, 80
Higham Special, 140
Highland Rally, 132
Hill, Phil, 97, 106

Hill, Walter, 87
Hobbs, David, 118–120
Hobbs, Howard, 118
Holt, Bill, 101, 107, 108
Hornburg, Charles, 106
Horning, 128, 129, 132
Howe, Lord, 19
Howorth, Hugh, 109
HRG, 81, 86
Humber, 14
HWM, 96

Ickx, Jacky, 130
Ickx, Jacques, 129, 130
Imhof, Godfrey, 129, 132
Indianapolis, 87
Institute of Mechanical Engineers, 23

Jabbeke, 26, 34, 46, 81, 122, 128, 135, 139, 140
Jaguar
  XJR, 13
  XK SS, 19
  1½ litre, 21, 44
  3½ litre, 44
  C-type, 57, 91, 103–107, 109–111, 128, 136, 140
  C/D hybrid, 140
  D-type, 19, 34, 57, 65, 69, 71, 75, 128, 138–141, 144
  E-type, 13, 14, 28, 34, 37, 41, 47, 51, 56, 62, 65, 115–117, 119, 120, 145, 149, 154, 168, 170
  E2A, 71
  LT cars, 106, 112, 113
  Mk I, 44, 61, 144, 180
  Mk II, 44, 75, 177
  Mk V, 26–28, 31, 42, 55
  Mk VII, 31, 67, 131, 132, 136, 138
  Mk X, 55
  XJ6, 44
  XK 100, 28, 30, 34
Jaguar Car Club, 170
Jaguar Drivers' Club, 116, 146, 169, 170
Jaguar World magazine, 169, 170
Jeep, 17
Jensen, 144
Johnson, Leslie, 80–85, 87, 90–97, 99, 100, 102, 127, 128, 133, 135
Johnston, Sherwood, 97
Jowett, 81, 82, 86, 92, 93, 95

Kettle, Roy, 111

Lagonda, 81, 86
Lakeland Rally, 127, 130
Lamborghini, Ferrucio, 9
Lancaster, 17
Lanchester, Fredrick, 57
Laroche, 130, 133
Lauren, Ralph, 116
Lawrie, Robert, 105
Le Mans, 13, 40, 71, 73, 80, 92–94, 97, 99, 102, 104–106, 111, 112, 118, 128, 137, 138, 140, 142, 168

Index

Lea Francis, 28
Lea, John, 91, 99, 112, 122
Liège–Rome–Liège Rally, 129, 130, 133, 138
Lisbon Rally, 138
Lister Jaguar, 118
Liston, Sonny, 96
Little Rally, 142
Loewy, Raymond, 11
Lola, 117
London Rally, 130, 131
Lyons–Charbonnières Rally, 136
Lyons, John, 129
Lyons, Sir William, 7–20, 26, 27, 31, 32, 34, 52, 57, 79, 80, 82, 87, 88, 92, 95, 102, 106, 122–125, 129, 130, 148, 169

MacDonald, David, 128
Mairesse, Willy, 92, 93
Mallory Park, 121
Mansbridge, Mr & Mrs, 137
Marshall, John, 92
Marzotto, Giannino, 89, 91
Maserati, 80, 90, 100
Massey, Rosemary, 121
Mathiesen, 138
Maudslay, Reginald, 12
Maximum velocity formula, 35
McKenzie, Bill, 129
McLaren, 37
Mena, Alfonso, 87
Mercedes-Benz, 57, 112
Meteor, 17
Meyrat, 92, 93
MG, 9, 81, 86, 93, 127, 168
Mille Miglia, 28, 87–91, 93, 99, 100
Miller, 40
Mini, 9, 145
MIRA, 140
Modena, 91
Monte Carlo Rally, 132, 136
Montlhéry, 128, 133, 134
Monza, 91
Moore, Oscar, 101
Morecambe Rally, 124, 128, 131, 132, 137
Morgan, 31, 86, 129
Morris, 9, 10, 17
Mosquito, 17
Moss, Alfred & Pat, 95
Moss, Stirling, 80, 83, 95–97, 99–102, 110, 111, 127, 128, 134, 135, 138, 169
The Motor, 33, 35
Motor Cycling Club, 131
Motor Sport, 29, 38, 100, 110
Murray, David, 112

Nissan, 149
Norton, 129
Nuvolari, Tazio, 94, 95, 102

Oblin, 130

Oldham & Crowther, 117, 118
Oporto, 92
Oulton Park, 114, 119, 130

Paget, Dorothy, 19
Palm Beach, Florida, 87
Paris–Dakar, 130
Parkinson, Don, 97
Parnell, Reg, 101
Parry Thomas, J.G., 140
Pearson, John, 116
Pearson, John N., 116, 117
Pebble Beach, 97
Pegaso, 139–141
Peignaux, Henri, 128, 136
Playford, Raymond, 169
Porsche, 137, 138, 141, 154
Portugal, 92
Potter, Len, 82
Preece, David, 117
Prescott, 105, 113, 120, 148
Protheroe, 114–116

RAC formula, 14
RAC Rally, 19, 131, 132, 136, 141, 143
RAC Tourist Trophy, 87
Radix, 130, 133
Railton, Reid, 15
Rainbow, Frank, 99, 100, 136
Rallye du Soleil, 128, 132
Ramponi, Giulio, 19
Renault, 75, 133
de Ridder, 140
Riley, 81, 93
Rolls-Royce, 30
Rolt, Tony, 82, 94–96, 101, 138, 140
Rosemeyer, Bernd, 102
Rosier, 93
Round Britain Rally, 142
Rover, 44
Royce, Henry, 9
Rutherford, Jack, 87

Salmson, 21
Salvadori, Roy, 103
Samworth, 129
Sayer, Malcolm, 104, 140
Schwelm, Adolfo, 94, 95
Scott, Denis, 138, 142
Scottish Rally, 132, 141, 142
Seaman, Dick, 19
Sears, Jack, 46
Shelsley Walsh, 39, 183
Sherpa van, 178
Silverstone, 32, 33, 35, 79–81, 85, 87, 92, 94, 95, 99, 100, 115, 121, 136, 138, 184
Sims, Lyndon, 144
Skilleter, Paul, 169
Smiley, Bob, 120
Snetterton, 46, 69, 114

191

Index

Snow, I.D., 132
Soler, 129
Speedometer calculation, 150
Spitfire, 17
Sports Car Club of America, 87, 121
SS, 90, 13, 14
   100, 14, 38, 42, 69, 81
  Car Club, 169
  Cars Ltd, 13, 17
SS 1, 12, 13
SS 2, 12
St John Horsfall, 'Jock', 80, 83
St Nicolas, 140
Standard, 11, 12
Standard Car Company, 10–12
Stewart, Jackie, 116
Stoss, Mr & Mrs, L., 142
Straight, Whitney, 19
Sturgess, Robin, 115
Sunbeam, 102
Sutcliffe, 129
Sutton, 'Soapy', 32
Sutton, Ron, 122, 136
Swallow, 8, 12
Swift, John, 108, 109

Talbot-Lago, 92–94
Targa Florio, 88, 89, 91
Taylor, Mr & Mrs, 133, 135
Thomson & Taylor, 15
Tour de France Automobile, 131

Tourist Trophy, 86, 95
Triumph, 19, 178
Tulip Rally, 123, 124, 128, 131, 144
TVR, 37
Tyrer, Gillie, 127

Uhlik, 120

Viking Rally, 138
Villoresi, Luigi, 89
Vivian, Charles, 142, 144
Volkswagen, 43

Walker, Peter, 80–85, 88, 92, 94, 95, 100, 101
Waller, Ivan, 105
Walmsley, William, 8, 13
Walshaw, Robert, 111
Weslake, Harry, 13, 23, 70
White Mouse racing team, 80
Whitehead, Peter, 80, 92, 95–97, 109
Whitley, 17
Wicken, George, 101, 102
Wilkins, Gordon, 130
Williams, Frank, 15
Wisdom, Bill, 14
Wisdom, Tommy, 14, 88, 90–92, 94–96, 99, 100
Wright, Joe, 67

XK 120 Specification, 36
  120 C Specification, 104
  140 Specification, 60